"Social history is, most elementally, food history. Jane Ziegelman had the great idea to zero in on one Lower East Side tenement building, and through it she has crafted a unique and aromatic narrative of New York's immigrant culture: with bread in the oven, steam rising from pots, and the family gathering round."

—Russell Shorto, author of *The Island at the Center of the World*

"An engaging and delicious slice of life on the Lower East Side. And the recipes found in this book, though originating from various cultures, all have the air of comfort foods and home."

—Joan Nathan, author of *Jewish Cooking in America*

"What do just-arrived immigrants see as they gaze around a new land, and what do their native-born neighbors see as the newcomers make their presence felt? More practically: How do people begin the work of putting food on their tables amid unfamiliar streets and languages? These questions couldn't be more timely. Nor could Jane Ziegelman's penetrating exploration of them. You will come away with a renewed sense of what it means to be an American."

—Anne Mendelson, author of *Milk* and *Stand Facing the Stove*

"A truly fine idea. It not only opens a window to view the ways in which our nation's immigrants cooked and ate, it broadens and enriches our understanding of the entire immigrant experience. This book is an impressive contribution to American cultural history."

—Nach Waxman, Kitchen Arts & Letters, New York City

"Jane Ziegelman brings us into the kitchens of five women whose home cooking not only fed their families and their neighborhoods but also became part of the culinary DNA of America itself. Drawing on wonderfully evocative primary sources, Ziegelman describes how they contributed to the complexities of ethnic identity, class, and religion in a tumultuous city. Beautifully written and full of insights, *97 Orchard* makes it clear that the story of New York is overwhelmingly a story about buying, selling, cooking, eating, and sharing food."

—Laura Shapiro, author of *Perfection Salad: Women and Cooking at the Turn of the Century*

"In this compelling foray into forensic gastronomy, Ziegelman pulls the facade off the titular 97 Orchard Street tenement. The result is a living dollhouse that invites us to gaze in from the sidewalk. With minds open and mouths agape, we witness the comings and goings of the building's inhabitants in the years surrounding the turn of the twentieth century. By focusing on the culinary lives of individuals from a variety of ethnic groups, Ziegelman pieces together a thorough sketch of Manhattan's Lower East Side at a time when these immigrants were at the forefront of a rapidly changing urban life. The food facts she uncovers are sure to interest and astound even those outside the culinary community, and guarantee that the reader will never look at a kosher dill pickle, a wrapped hard candy, or even the delectable foie gras the same way again. Ziegelman cleverly takes this opportunity to show us that in learning about food, we're actually learning about history—and when it comes to the sometimes surprising journey some of our favorite meals have taken to get here, it's fascinating stuff."   —*Booklist*

"This whole book is a celebration of food, language, and of the mutual aid and comfort that these brave pioneers shared with their tenement neighbors and the citizens who took them in."

—Julie Wittes Schlack, *Boston Globe*

"Blending history, sociology, anthropology, and economics, spiced with recipes, Ziegelman offers a look at the Lower East Side and the immigrants who made it legendary." —*Chicago Jewish Star*

"It is an eye-opening exploration of the social and economic history of those who thrived and survived, in spite of significant odds, on New York's Lower East Side. VERDICT: Recommended for those seeking up-close and personal—as well as edible—insights into the daily lives of late nineteenth- and early twentieth-century 'new Americans.'" —*Library Journal*

"A tasty, satisfying stew of history, sociology, cultural anthropology, and spicy prose." —*Kirkus Reviews* (starred review)

"A welcome addition to the canon inspired by the Tenement Museum on the Lower East Side. . . . [Ziegelman] dishes delectable morsels of ethnic gastronomy." —*New York Times*

"Ziegelman puts a historical spin to the notion that you are what you eat. . . . Ziegelman vividly renders a proud, diverse community learning to be American. Through food, the author records the immigrants' struggle to reinterpret themselves in an American context and their reciprocal impact on American culture at large." —*Publishers Weekly*

"In Ziegelman's telling, 97 Orchard and places like it become sites of culinary tenacity and cultural creativity whose contributions to the American scene need to be remembered. . . . Ziegelman shows how immigrants contributed to the U.S. diet not only as home cooks and cookbook writers but also as bakers, brewers, butchers, and restaurateurs. . . . Readers will find much to savor in *97 Orchard*." —*Christian Century*

"Beautifully written and thoroughly researched, *97 Orchard* brings this world vividly to life as it describes how America transformed the waves of late nineteenth-century immigrants and how they changed America. . . . Ziegelman weaves a rich, frequently astonishing social tapestry of the period—ranging from discussions of real estate and theater to include dozens of popular recipes."

—Gloria Levitas, *Moment*

"Highly entertaining and deceptively ambitious, the book resurrects the juicy details of breakfast, lunch, and dinner (recipes included) consumed by poor and working-class New Yorkers a century and more ago. . . . Ziegelman adroitly works her way through the decades and her five cuisines."   —*New York Times Book Review*

"Ziegelman is a patient scholar and a graceful writer, and she rummages in these families' histories and larders to smart, chewy effect."

—Dwight Garner, *New York Times* (Books of the Times)

"A satisfying stew to be savored after one's favorite American repast."

—*USA Today*

"*97 Orchard* offers an eminently readable history of five families. . . . But the book's beauty lies in the insight and intelligence with which Ziegelman tells the story of real, live people who came to this country and brought their food with them. Using census records, shopping lists, recipes, and other documents, she brings her characters to life, and illuminates how immigrant food such as hot dogs and pizza became American food. . . . A must-read for anyone interested in food, ethnicity, and culture."   —*Winston-Salem Journal*

97 ORCHARD

ALSO BY JANE ZIEGELMAN

Foie Gras: A Passion

97 ORCHARD

An Edible History
—
of Five Immigrant Families
—
in One New York Tenement

JANE ZIEGELMAN

HARPER

NEW YORK · LONDON · TORONTO · SYDNEY

For Andy

HARPER

A hardcover edition of this book was published in 2010 by Harper, an imprint of HarperCollins Publishers.

97 ORCHARD. Copyright © 2010 by Jane Ziegelman. All rights reserved. Printed in the United States of America. No part of this book may be used or reproduced in any manner whatsoever without written permission except in the case of brief quotations embodied in critical articles and reviews. For information address HarperCollins Publishers, 10 East 53rd Street, New York, NY 10022.

HarperCollins books may be purchased for educational, business, or sales promotional use. For information please write: Special Markets Department, HarperCollins Publishers, 10 East 53rd Street, New York, NY 10022.

FIRST HARPER PAPERBACK PUBLISHED 2011.

Designed by Mary Austin Speaker

The Library of Congress has catalogued the hardcover edition as follows:

Ziegelman, Jane.

　97 Orchard : an edible history of five immigrant families in one New York tenement / by Jane Ziegelman. — 1st ed.

　　p. cm.

　　Includes bibliographical references and index.

ISBN 978-0-06-128850-0 (hardback)

　1. Food habits—New York (State)—New York—History—19th century. 2. Immigrants—Nutrition—New York (State)—New York—History—19th century. 3. Lower East Side (New York, N.Y.)—History—19th century. 4. Lower East Side (New York, N.Y.)—Social life and customs. I. Title.

GT2853.U5Z54 2010

394. I'20974741—dc22　　　　　　　　　　　　　　　　　　2009049637

ISBN 978-0-06-128851-7 (pbk.)

12 13 14 15 ᴏᴠ/ʀʀᴅ 10 9 8 7 6 5

Contents

———

Acknowledgments

This book would have no reason to exist if not for the Lower East Side Tenement Museum, the present-day 97 Orchard Street. I am forever indebted to Ruth Abram, founder of the museum and the woman who granted this project the spark of life. I also need to thank Morris Vogel and Helene Silver for their steadfast support, and David Favaloro and Derya Golpinar for sharing their time and their knowledge.

In the course of researching this book I have benefited from the guidance of a small army of food authorities, genealogists, historians, and librarians. I would like to thank Karen Franklin, Roger Lustig, Joel Hecker, Lori Lefkowitz, Vivian Ehrlich, Anne Mendelson, Joan Nathan, Lorie Conway, Roberta Saltzman, Eleanor Yadin, Amanda Siegel, Bonnie Slotnik, Barry Moreno, and Janet Levine. I am likewise grateful to the immigrants, their children, and grandchildren who shared their stories and their recipes. Among them are Barbara Levasseur, Flora Frank, Brian Biller, Josef Griliches, Hannah and Walter Hess, Maria Capio, Francine Herbitter, Lillian Chanales, Betsy Chanales, Frieda Schwartz, and Edy Geikert. And of course, I must thank my incredibly patient editor, Elisabeth Dyssegaard, and my agent, Jason Yarn. Finally, I would like to thank Marjorie and Aaron Ziegelman, Michael Coe, and my friends Stephen Treffinger, Steve Miller, and Joshua Patner for being such perceptive and tireless readers.

Introduction

97 *Orchard* tells the story of five immigrant families, each of them, as it happens, residents of a single New York tenement in the years between 1863 and 1935. Though separated by time and national background, the Glockners, the Moores, the Gumpertzes, the Rogarshevskys, and the Baldizzis, were all players in the Age of Migration, a period of sweeping demographic change for both the Old and New Worlds.

Starting in Europe in the early 1800s, whole chunks of humanity streamed from the countryside to the cities—the continent's new manufacturing centers—in pursuit of work. Those who could afford to embarked on a trans-Atlantic migration, lured to the United States by the promise of American prosperity and freedom. *97 Orchard* chronicles what became of those immigrants, but from a special vantage point: it retells the immigrant story from the elemental perspective of the foods they ate.

Within hours of landing, immigrants felt the keen pressures of assimilation. Before they even left Ellis Island, many had already traded in their Old World identities for new American names. Once on the mainland, immigrants found it expedient to shed their native clothing and to dress like Americans. Men quickly adopted the ubiquitous derby. Women abandoned their shawls and kerchiefs in favor of American-style coats and bonnets. The immigrants learned to speak like Americans, subjected themselves to the rigors of American sweatshops, and delighted in the popular culture of their adopted home. These same immigrants, however, went to extraordinary lengths to preserve their traditional foods and food customs. Transplanting Old World food traditions—many of them rooted in the countryside—to the heart of urban America required both imagination and tenacity. To compound the challenge, the immigrants' eating habits oftentimes defied American culinary norms, and as the immigrant population continued to swell, concerned citizens attempted to wean the foreigners from their strange cuisine. The immigrants' food loyalties, however, were fierce. Native foods provided them with the comfort of the familiar in an alien environment, a form of emotional ballast for the uprooted. Within the immigrant community, food cemented relationships, and immigrants turned to food as a source of ethnic or national pride. As immigrant families put down roots, it also became a source of contention between parents and their American-born children for whom Old World foods carried the stigma of foreignness.

A large part of this story takes place in the immigrant kitchen. For many immigrants, this was a small, often windowless room in a five- or six-story brick tenement. A form of urban housing that began to appear on New York's Lower East Side in the 1840s, tenements were the first American residences built expressly for multiple families—in this case, working people. The typical tenement had an iron front stoop, a central stairwell, where children played and neighbors socialized, and four apartments on every floor. The tenement kitchen was furnished with a wood- or coal-burning stove and little else. Those at 97 Orchard, a well-equipped building for its time, were bereft of indoor plumbing or any means of cold storage aside from the windowsill or fire escape, a makeshift "ice box" that only functioned in winter. A place to cook and to eat, the kitchen was also used as a family workspace, a sweatshop, a laundry room, a place to wash one's body, a nursery for the babies, and a bedroom for boarders. In this cramped and primitive setting, immigrant cooks brought their formidable ingenuity to the daily challenge of feeding their families. 97 Orchard describes exactly how that challenge was met by five major immigrant groups: the Germans, Irish, German Jews, Russian-Lithuanian Jews, and Italians.

East Side children were responsible for collecting wood and coal for the family stove.

To procure the ingredients they needed at prices they could afford, immigrant cooks depended on neighborhood food purveyors. Upon landing in America, immigrant entrepreneurs quickly established networks of food laborers, tradespeople, importers, peddlers, merchants, and restaurant-keepers. Many of these culinary workers have since vanished and are long-forgotten. Among the disappeared are the German *krauthobblers*, or "cabbage-shavers," itinerant tradesmen who went door to door slicing cabbage for homemade sauerkraut; the Italian dandelion pickers, women who scoured New York's vacant lots for wild salad greens; and the urban goose-farmers, Eastern European Jews who raised poultry in tenement yards, basements, and hallways.

The networks they established met the foreigners' own culinary needs, but in the process of feeding themselves, they revolutionized how the rest of America ate.

A time traveler to pre–Civil War New York or Boston or Philadelphia, who happened to arrive at dinner time, could expect to encounter the following on the family table: roast beef stuffed with bread crumbs and suet, a dish of peas, and some form of pudding. This was sustenance for the professional or business class. Further down the economic ladder, generations of working-class Americans survived on "hash," a composite of leftover meat scraps and potatoes. One food that united the "haves" and "have-nots" was pie. Apple pie, cherry pie, berry pie, lemon pie, and mince pie were eaten for breakfast, lunch, dinner, and dessert. The habit was so pronounced that immigrants referred to their American hosts as "pie-eaters." Another universal food was oysters. While Americans devised a wealth of oyster-based recipes, including oyster patties and stews, they enjoyed them best in their natural state, sold raw from the saloons and street stands that proliferated in nineteenth-century cities.

The immigrants that began to settle in the United States in the 1840s introduced Americans to an array of curious edibles beyond their familiar staples: German wursts and pretzels, doughnut-shaped rolls from Eastern Europe known as "beygals," potato pastries referred to as "knishes," and the elongated Italian noodles for which Americans had no name but came to know as spaghetti. 97 Orchard describes how native-born Americans, wary of foreigners and their strange eating habits, pushed aside their culinary (and other) prejudices to sample these novel foods and eventually to claim them as their own.

Aside from satisfying our culinary curiosity, the exploration of food traditions brings us eye to eye with the immigrants themselves. It grants us access to the cavernous beer gardens that once lined the Bowery, where entire German families—babies included—spent their Sundays, the immigrant's only day of leisure, over mugs of lager beer and plates of black bread with herring. It is a door into the East Side cafés where Jewish pushcart peddlers drank endless cups of hot tea with lemon, accompanied by a plate of blintzes, and brings us face-to-face with the Italian laborers who formed their own all-male cooking communities to satisfy their longing for macaroni.

On the streets of the Lower East Side, European food customs collided with the driving energy of the American marketplace. The tantalizing saga that ensued, an ongoing tug of war between culinary tradition and American opportunity, goes to the heart of our collective identity as a country of immigrants. But while *97 Orchard* is concerned largely with a single immigrant community, Manhattan's Lower East Side, the history it tells transcends that one urban neighborhood. Though on a smaller scale, comparable changes were underway in cities and towns across America wherever immigrants settled. In fact, though the actors have changed, the culinary revolution that began in the nineteenth century continues today among immigrants from Asia, Africa, the Caribbean, and Latin America, who have brought their food traditions to this country and continue to transform the way America eats.

CHAPTER ONE

The Glockner Family

The Lower East Side of Manhattan, circa 1863, was a neighborhood of squat wooden row houses, shelter for a population of artisans, laboring people, and small-time tradesmen. Built decades earlier as single-family homes, by the time of the Civil War the ground floor of the typical East Side dwelling was generally taken up by a grog shop or grocery with a small apartment behind the store for the shopkeeper's family. Two more families lived on the second floor, while the basement was rented out to lodgers. More imposing structures could be found on the neighborhood's oldest streets. Made of stone, with peaked tile roofs, these were the former homes of New York's merchant princes, now converted into boardinghouses and cheap hotels that catered to a mainly immigrant clientele. But the East Side was also home to a strictly modern form of urban housing: the tenement—a five- or six-story brick building with multiple apartments on every floor. Their massive size, along with their plain facades, reminded nineteenth-century New Yorkers of army barracks, and they were often referred to that way, even by the people who lived in them.

Hidden behind the dwellings, in the shadowy courtyards within each city block, were machine shops, print shops, brick-makers, furniture and piano factories, to name just a few of the local industries. Another kind of factory was concealed within the tenement itself. Here, in apartments that doubled as sweatshops (a term that had not yet been coined), immigrant workers produced clothing, lace, cigars, and artificial flowers for ladies' bonnets, a valued commodity in the hat-wearing culture of the nineteenth century. More evident to the casual observer, however, was the neighborhood's vibrant commercial life. In other parts of the city, people lived in private homes on relatively quiet residential streets but shopped and caroused on the noisier, more bustling avenues. On the Lower East Side, that distinction was blurred. Some kind of shop or business occupied the street level of most East Side buildings, turning the neighborhood into a single teeming marketplace. East Side shops sold a vast array of goods, from rusted scrap metal and secondhand corsets to peacock feathers and beaver-skin coats. There were shoe and hat shops, apothecaries, blacksmiths, glaziers, and tailors. Most plentiful, however, were businesses related to food. The impressive concentration of food markets and food peddlers, of slaughterhouses, brewers, bakers, saloons, and beer halls satisfied the culinary needs of the immediate neighborhood. At the same time, they played an essential role in feeding the larger city.

The people who lived and worked on the Lower East Side were predominately immigrants and, in lesser numbers, people of color—freed slaves and the descendants of slaves. Those sections of the Lower East Side that had been settled chiefly by Germans were collectively known as *Kleindeutschland,* or "Little Germany," covering the area from 14th Street south to Division Street and from the Bowery all the way east to the river. The businesses here were German-owned; the newsboys hawked German-language newspapers, and the corner markets sold loaves of molasses-colored pumpernickel and rosy-pink Westphalian hams. This semi-discrete corner of New York, a city within a city, was the world inhabited by Lucas Glockner, his wife, Wilhelmina, and their five children. It is also the world we are about to enter.

But before we do, let's have Mr. Glockner say a few words on his own behalf. Dead now for over a century, he speaks to us nonetheless with the

*1870 census record for Lucas Glockner and his family. Census
records, among other official documents, provide valuable information
on the lives of otherwise anonymous immigrants.*

help of certain official documents, key among them the federal census
report. The first census in which his name appears was taken in 1850,
roughly four years after Glockner's arrival in New York. While the United
States government had been counting its citizens since 1790, the 1850
census was groundbreaking in one respect: for the first time, it recorded

the names of all household members, including women, servants, slaves, and children. Because of this innovation, we know that in 1850, Mr. Glockner lived on the Lower East Side at 118 Essex Street, along with his first wife, Caroline, a four-year-old son named Edward, and a baby named George, who was one at the time and would not survive. In this document, Mr. Glockner describes himself as a tailor, the leading occupation among New York Germans. According to the 1850 census, he is one of seven tailors, all of them German, living in the same small building.

The next time we hear from him, the United States is locked in a bloody civil war, and Lucas Glockner, along with thousands of other East Side Germans, has been registered to serve in the Union Army. According to an 1864 draft record, a beautiful, hand-lettered document, he is still employed as a tailor. Other sources tell us, however, that Glockner is ready to abandon tailoring for the more lucrative career of a New York property owner. In fact, he has already made his first investment. Glockner and his two partners have pooled their money to buy up the Dutch Reformed Presbyterian Church, not for the building but for the land underneath it: a plot large enough to fit three typical East Side tenement buildings. By the time of the next census in 1870, Glockner has become a rent-collecting landlord, the owner of several East Side properties.

By 1880, Glockner is living at 25 Allen Street with his considerably younger wife, Wilhelmina. Together they have three children: Ida, Minnie, and William. Neither of the girls is attending school, which shouldn't surprise us. If they weren't earning money as seamstresses or flower-makers, East Side girls were generally kept at home to help with the unpaid business of housework. Fifteen-year-old William, on the other hand, is enrolled in college, a very good indication that he will go on to work in an office—as a clerk, perhaps, or a bookkeeper, the kind of job that immigrant parents dreamed of for their sons. And Mr. Glockner? Living comfortably off his various properties (he owned at least three buildings by this time), he has earned the right to a new job title. At fifty-nine years old, Glockner describes himself as a "Gentleman." And there we have it, from tailor to gentleman, the basic trajectory of one human life. Mr. Glockner's autobiography.

Glockner earned his fortune by investing in the kind of buildings he knew best, the multifamily dwellings known as "tenant houses," or "tenements" for short. His first property was 97 Orchard Street, the five-story brick structure that stands at the core of our story. Built by Glockner on the grounds of the old Dutch Church, it was a compact building designed to maximize space, the mandate behind all tenement architecture. Covering a scant three hundred and fifty square feet, the Orchard Street apartments were minuscule by today's standards, the largest room not much bigger than a New York taxi. And yet, Glockner's building had a sense of style about it, both inside and out, a break from the tenement tradition up to that time.

Tenements, loosely defined, began to appear in New York sometime in the 1820s, many of them clustered in the old Five Points, a section of the Lower East Side that is now part of Chinatown. In colonial times, that same patch of New York had been a semi-industrial area of slaughterhouses, tanneries, breweries, rope- and candle-makers, all centered around a five-acre pond known as the Collect. In the early 1800s, the Collect was drained and filled, though not very effectively. A neighborhood of wood-frame row houses grew up on the site, but after a good hard rain, foul-smelling muck would well up from the ground, as if the former pond was reclaiming its rightful place. The terrible stench, along with the fear of disease, pushed out the old inhabitants, the merchants, and the craftsmen, making way for a less privileged class of day laborers, boot blacks, and laundresses. Desperate for shelter, they moved into old single-family homes, which had been carved up into apartments. These improvised structures were the city's original tenements.

The appearance of the tenement coincided exactly with a sharp rise in immigration that began in the 1820s, gathering momentum in the 1830s and 1840s. In its wake, the population of New York suddenly ballooned, creating the city's first housing crisis. City landlords quickly grasped how to profit from the situation. They bought up old houses, stables, and workshops, or converted buildings they already owned, dividing them up into cubbyhole-sized living quarters. For businessmen of the time, including John Jacob Astor, a major investor in the East Side housing boom, the tenement was a real estate windfall. Among the first purpose-fully built tenements was a five-story brick structure on Water Street, near the East River, financed by a New York businessman named James Allaire, owner of the Allaire Iron Works, a company that made steamship engines. Since nineteenth-century employers often supplied their workers with room and board, it seems a good possibility that Allaire's tenement was built for his employees.

The history behind 97 Orchard sets it apart from the investments of the Astors and Allaires of New York. Where most East Side developers were "building down," creating housing for people far beneath them in the social hierarchy, 97 Orchard was built by an East Side immigrant for people much like himself. In fact, Glockner and his family lived at 97 for the first half dozen years of the building's existence and remained tied to it through a web of personal relationships long after they moved. The Glockners had friends at 97, like Natalie Gumpertz, the German dress-maker abandoned by her husband, and John Schneider, who ran a saloon in the building's basement. More personal still, one of Glockner's sons eventually married the daughter of an Orchard Street tenant and moved into the building with his new wife.

The red-brick facade of 97 Orchard is an example of nineteenth-century Italianate design, very much in fashion during the 1860s. Typical of an Italianate row house, the kind seen farther uptown, the doorway at 97 Orchard is framed by a stone arch. Curved lintels and a stone sill border the windows, while the roof line is defined by a surprisingly ornate cornice. Though made of cast metal, it was finished to resemble brownstone, a more expensive building material. In fact, all of the build-ing's decorative elements were much simplified, discount versions of their uptown counterparts, the best that Glockner could afford. The basement at 97, which sits just below street level, is occupied by stores, one on either side of the building's front stoop.

On climbing the stoop, one enters the residential part of the build-ing. The first room is a vestibule, or entryway, the walls lined with panels of white marble. On the far side of the vestibule door, a narrow hallway leads to a plaster arch. Passing under it, the hallway widens.

Directly ahead is the heavy wooden stairway that runs up the center of the building.

The apartments at 97 Orchard comprise three small rooms, a parlor, a kitchen, and a windowless "dark room" used for sleeping. Despite their size, the rooms are smartly finished with light oak baseboards and chair rails that match the doors and window frames. The walls are painted in pastel shades like salmon pink and pale mint green, while the ceilings are painted a soft shade of sky blue. Each apartment has two fireplaces, one in the kitchen used for cooking, and another in the parlor with a wooden mantel and slate hearthstone.

It had taken Glockner years of saving to buy the Orchard Street real estate and put up his building, a huge investment for an immigrant tailor, and a huge risk as well. Though he still had his trade, all of his capital was now in the building, a precarious state of affairs for a man in his forties with a family to support. Despite all this, Glockner embellished his property with marble paneling, arched doorways, chair rails, fireplaces with proper mantels. All of these flourishes are representative of Glockner's attempt to reach beyond *Kleindeutschland* and participate in the larger and more affluent culture of middle-class New York.

Though he splurged on décor, he skimped in other ways. Of all his money-saving strategies, none was more glaring than the absence of indoor plumbing. By 1863, pipes carrying fresh water from the Croton aqueduct had been laid under Orchard Street, and Glockner could have easily tapped into the underground system. Instead, he provided the building with a row of privies and an outdoor pump, both located in the building's back courtyard. Everyone who lived at 97 felt the impact of Glockner's decision, but no one felt it more than the building's women. Tenement housewives were like human freight elevators, hauling groceries, coal, firewood, and children up and down endless flights of stairs. Their most burdensome loads, however, were the tubs of water needed for laundry, bathing, house-cleaning, and cooking. It was sloppy, muscle-straining work, water sloshing everywhere, soaking the stairs and the women too, a bone-chilling prospect on a cold February morning, especially since the stairs were unheated.

Once a week the tenement kitchen served as a laundry room. Women and girls were responsible for hauling water up and down the stairs.

The premium on water shaped the way women cooked in the tenements. Climbing up and down three or four flights of stairs just to wash a dish is strong motivation to cook as simply and efficiently as possible. Lucky for Mrs. Glockner, Germans were expert stew-makers, a very useful culinary skill since it provided an entire meal using a single pot. Following German tradition, lunch was the heartiest meal of the day in the Glockner apartment. In the evening, the family might have boiled eggs, or bread and cheese, but lunch was a time to feast, time to fill your stomach with a good German fricassee of beef or veal or pork, served with boiled dumplings or maybe noodles.

Imagine, for a moment, a typical morning in the Glockner household. Mrs. Glockner is out, shopping for groceries, the baby is upstairs with a neighbor, so Mr. Glockner can attend to his accounts. At a small table by the parlor window, bent over his ledger, twirling the end of his rather bushy mustache, he loses himself in the rows of numbers. Very satisfying, he thinks, to see them all lined up so neatly. (After decades as a tailor, he appreciates good craftsmanship.) His thoughts are interrupted by the return of his wife. Hanging her cloak on a brass hook next to the door, she gives her hands a brisk rub to get the circulation back (the fall weather has suddenly turned cold) and lights a fire in the new black stove. Now she turns her attention to fixing the stew. The smell of browning onions reminds her husband that it's time for his mid-morning snack, so he trots downstairs to Schneider's Saloon, conveniently located in the basement of the building, for a quick pint of beer and a plate of herring. He spends an hour or so chatting with Schneider, by which time the stew is nearly ready.

Recipes for German stews of the period can be found in the *Praktisches Kochbuch* (*Practical Cookbook*), by Henrietta Davidis, Germany's answer to Fanny Farmer. Originally published in Germany in 1845, the *Praktisches Kochbuch* offers a sweeping view of what Germans were eating in the nineteenth century. The book was tremendously popular, selling over 240,000 copies in the author's lifetime. Some of those copies traveled to America in immigrant suitcases. Additional copies were shipped across the Atlantic and sold in German-language bookstores in the United States. In 1879, a German-bookstore owner in Milwaukee, a city with a large German community, published the first American edition of Henrietta Davidis under the title *Praktisches Kochbuch fur die Deutschen in Amerika* (or *Practical Cookbook for Germans in America*). A bestseller in immigrant circles, the book was reprinted several times. The first English translation, which appeared in 1897, reached a different and wider audience. It was for the immigrants' children and grandchildren, people who spoke English as their first language, and who had perhaps lost touch with the cooking traditions of their German ancestors. But the book also appealed to ordinary Americans of any background, since by 1897, many had sampled German cooking and wanted to know more.

The *Practical Cookbook* contains many stew-like recipes, some called "fricassees" and some "ragouts," but all of them savory concoctions of meat, vegetables, broth, and assorted seasonings. A recipe for stewed duck with dumpling uses pork fat, peppercorns, cloves, bay leaves, onion, and lemon peel to flavor the cooking stock. A recipe for stewed leg of mutton calls for the meat, which "should not be too fresh," to be simmered in water and beer, and seasoned with "cloves, peppercorns, three bay leaves, a few whole onions, and a bunch of green herbs, such as garden rue, marjoram, and sweet basil."[1] The generous use of spices and fresh herbs, the hint of tartness from lemon or vinegar, make all these dishes typically German. But for even more concentrated flavor, the *Practical Cookbook* provides a recipe for spiced vinegar, a condiment for sprinkling over stews at the table, like a German form of Tabasco sauce. The potent mixture calls for a "half ounce of mace, some cloves (or, if preferred, garlic), ginger, one ounce of mustard seed, a pinch of whole white pepper, a piece of grated horseradish, a handful of salt, six or eight bay leaves," all steeped in a jar of vinegar along with sixty whole walnuts.[2]

In the world of German stews, perhaps no dish was more highly flavored than *hasenpfeffer*, a ragout made from wild rabbit. Immigrants brought their love for *hasenpfeffer* to New York, where German saloon-keepers gave bowls of it to anyone who paid for a drink. Below is a recipe for *hasenpfeffer* by Gesine Lemcke, a German immigrant who opened a successful cooking school on Manhattan's Union Square. She also wrote cooking columns for the *Brooklyn Eagle*, which is where this recipe appeared in 1899:

HASENPFEFFER

Cut two well-cleaned rabbits into pieces, season them with one tablespoonful salt, put them in a bowl, add two large onions cut in slices, six cloves, twelve whole allspices, and half tablespoonful whole peppers, cover with vinegar; cover the bowl and let stand three days. When ready to cook, put the rabbits with the vinegar and all the other ingredients into a saucepan over the fire, add half pint water and table-spoonful sugar, boil till tender. In the meantime, melt one heaping tablespoonful butter, add one heaping tablespoon flour, stir until light brown, strain the rabbit broth, add it to the flour and butter, stir and cook to a smooth creamy sauce, lay the rabbit in a hot dish and pour the sauce over it. Serve with small browned potatoes cooked in deep fat or serve with potato dumplings.[3]

The following is a recipe for German-style veal stew with celery root and dried pear. Lemon, mace, clove, and bay leaves are the main seasonings, a combination often found in German-American cookbooks of the period. Bringing together meat, root vegetables, and fruit is another common German touch.

Veal Stew with Dried Pear

2 ½ pounds veal stew meat
½ pound veal or beef bones
3 tablespoons butter
I large onion, chopped
½ cup chopped parsley root
I rounded tablespoon flour
small pinch of mace (about ⅛ teaspoon)
2 whole cloves
I ½–2 cups beef stock
6 stalks parsley
8 dried pear halves, cut in half lengthwise
I medium celery root, peeled, cut in half then thinly sliced
½ lemon, thinly sliced into rounds

Rinse the meat and pat dry. In a large Dutch oven or heavy stew pot, melt 2 tablespoons of the butter. When it begins to foam, add veal in batches. Be careful not to crowd the pot or the meat won't brown properly. Let the veal cook, untouched, five minutes or so before turning it to brown the

other side. You should also brown the bones. Remove veal,
bones and all, from the pot. To the same pot, add onion and
parsley root. Sauté until golden, adding more butter if needed. Add the flour, and stir for a minute or so. Return veal to
the pot, seasoning it with salt and pepper, mace and cloves.
Add the beef stock, just enough to cover, along with bay
leaves, 6 stalks parsley, and dried pear. Simmer very gently
for about 1 ½ hours. Add celery root and cook another half
hour. In the last ten minutes, add the sliced lemon.[4]

In German kitchens, the traditional accompaniments to stew were some
form of dumplings. Bread dumplings, potato dumplings, flour dumplings, dumplings made with cabbage, bacon, liver, ham, sweetbreads, or
even calf's brain—these are just a few of the dumpling recipes found in
early German-American cookbooks. In most volumes, an entire chapter
is given over to them, both savory and sweet. The following bread-based
recipe for "Green Dumplings," flecked with bits of chopped parsley,
spinach, and chive, is from Henrietta Davidis:

Green Dumplings (a Suabian [sic] recipe)

A handful of parsley, the same quantity of spinach, half as
much chervil and chives, chop all together and stew in butter for a few moments. Then mix with 2 grated rolls, 2 eggs,
salt and pepper, form into little balls, and let them come
just to a boil in the finished soup, or they will fall to pieces.
These dumplings are very nice in the Spring.[5]

The alternative to dumplings was noodles, a Bavarian specialty that German cooks adapted from the Italians, their neighbors to the south. In
the German state of Swabia, cooks perfected a technique for making
the pebble-shaped noodle known as spaetzle. Bavarians made threadlike

soup noodles and thick, chewy noodles eaten as a side dish. Here is Ms. Lemcke's recipe from 1899:

Egg Noodles

Put one cup of flour in a bowl, add two eggs, a small piece of butter the size of a hazelnut, a pinch of salt, and two tablespoonfuls cold water, mix this into a dough, adding more flour if necessary, turn the dough onto a board and work it till stiff and smooth, divide it into four parts, roll each part out very thin, hang them over the edge of a bowl to dry, then roll each piece up like a music roll and if the *nudels* are wanted for soup cut them as fine as possible, and if wanted to be served with fricassee in place of vegetables cut them half-finger wide. As soon as they are cut, shake them apart on a floured board and let them lie until perfectly dry.[6]

In 1865, Orchard Street alone was home to at least ten grocery stores, most of them German-owned. Ten years earlier, the same stores would have been in Irish hands. As German immigrants flowed into the city in the 1850s, the balance began to shift. By 1860, the German corner grocery had become a New York fixture, not just on the Lower East Side but throughout Manhattan and Brooklyn too.

The trick to running a successful grocery was to "have a little bit of everything and no great quantity of anything." These stores typically carried a small selection of fruits and vegetables, milk and butter, canned goods, coal, kerosene, kindling wood, sugar, soap, rolled oats, crackers, cigars, and for their German customers, imported delicacies like Westphalian ham, caviar, sausages, sauerkraut, and poppy-seed oil. But their best-selling item was alcohol, usually whiskey, which provided the grocer with most of his income. As it happens, 97 Orchard Street was literally flanked by grocers. If Mrs. Glockner needed a cup of milk or a loaf of bread, she could dash downstairs and buy it from either Frederick Aller at number 95 or Christian Munch at 99. Most likely, she bought on credit, the normal way of doing business for a grocer of the period. At the end of the business week, her husband would drop by the store and pay the bill.

For more serious shopping, Mrs. Glockner hooked her basket over one arm and headed for the public market on Grand Street, one of roughly a dozen scattered through Lower Manhattan. The public markets were large, shedlike structures with rows of individual stalls, the largest by far the Washington Market on the Lower West Side. Conveniently located near the busy docks along the Hudson River, this was the place where most of the food consumed in New York was bought and sold.

A quick scan of city newspapers circa 1860 reveals how much negative attention was generated by the public markets. The main complaint: dirt. The following "warning" ran in the *New York Times* in May 1854:

> *If you are going to market this morning, be pleased to put on thick, stout shoes, and a dress that will not readily show dirt. For of all the dirty places in the City, our Public Markets are the dirtiest. In the fish markets the floors are slippery and constantly wet. In the meat market, giblets are scattered about the floor, unsightly objects are obtruded at all points, and refuse meats are frequently only swept out under the eaves, and left to disgust all passersby.*[7]

Aside from the filth, the condition of the buildings themselves, patched together and half-disintegrating, was deplorable. Among the most decrepit was the Fulton Market, "a filthy wood-shed with its leaky roof and tottering chimneys."[8] For observers of the time, it was hard to reconcile the dirt and decay of the markets with the stature of New York, the largest, richest city in America. "The Metropolitan city of New York has endured the stigma of being, without question, the most illy-supplied with public food markets of any civilized centre of population of even one-tenth its pretensions," is how one critic put it.[9]

The attitude of shamed outrage was just about universal, but not quite. The many accusations hurled at the markets belied an immutable fact, one recognized by a select handful of supporters. The market system supplied New Yorkers with a staggering variety of meats, fish, fowl, vegetables, and fruits. The following description comes from *The Great Metropolis*, Junius Henri Browne's 1869 guide to New York. At the Washington Market, he tells us,

> *Nothing is lacking to gratify the palate,—to delight the most jaded appetite. The best beef, mutton, veal and lamb the country affords are displayed upon the stalls. Those roasts and steaks, those hind-quarters, those cutlets, those breasts with luscious sweetbreads, would make an Englishman hungry as he rose from the table. Those delicate bits, so suggestive of soups, would moisten the mouth of a Frenchman. Those piles of rich juicy meats would render an Irishman jubilant over the memory of his determination to emigrate to a land where potatoes were not the chief article of food. What an exhibition of shell-fish, too! Crabs, and lobsters, and oysters in pyramids, yet dripping with sea-water, and the memories of their ocean-bowers fresh about them. And vegetables of every kind, and fruits, foreign and domestic, from the rarest to the commonest, from the melon to the strawberry, from the pine apple to the plum. Fish from the river and mountain stream, from the sea and the lake. Fowls and game of all varieties, from barnyard and marsh, forest and prairie.*[10]

Critics of the public market took for granted the feast available to them on a daily basis; they were equally blasé about the tremendous human effort required to assemble all those varied goods: beef and pork transported by rail from the Midwest; vegetables, butter, cheese, and milk from the farms of Connecticut, New Jersey, and Long Island; stone fruits and melons from the South, along with fish and seafood shipped from all points along the Eastern Seaboard.

A leading defender of the markets was Thomas De Voe, a New York butcher who leased a stall in the Jefferson Market at the intersection of Sixth Avenue and Greenwich Street. A portrait of De Voe shows him in typical butcher's costume: a top hat and long apron, a knife in one hand, poised before a rack of meat, ready to slice.

Born in 1811, De Voe worked as a butcher's apprentice as a young boy and remained with the profession until 1872, the year he was appointed superintendent of markets for the city of New York. But De Voe was an intellectual as well, intensely curious about the world of the market and how it evolved. In 1858, he presented a paper on the history of the

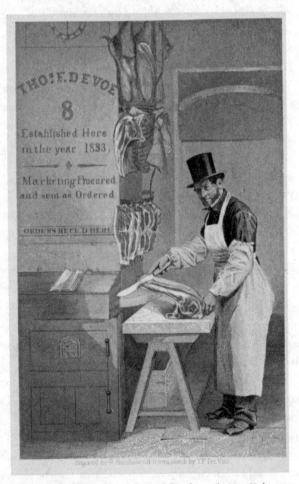

Portrait of Thomas De Voe, scholar and defender of the New York public markets.

markets to the New-York Historical Society, which he later expanded and published as *The Market Book*. His next project, *The Market Assistant*, was an encyclopedic and exhaustively researched survey of "every article of human food sold in the public markets of New York, Boston, Philadelphia, and Brooklyn."[11] The result of his efforts is a precise record of culinary consumption in urban America. It tells us, for example, that New Yorkers once dined on buffalo, bear, venison, moose (the snout was especially delectable), otter, swan, grouse, and dozens of other species, wild and domestic; that fish dealers offered fifteen types of bass, six types of flounder, and seventeen types of perch; and that shoppers at the produce stalls could choose between purslane, salsify, borage, burdock, beach plum, black currants, mulberries, nanny berries, black gumberries, and whortleberries.

Business at the public markets followed a predictable daily rhythm. It began at four in the morning, when the wholesale customers—the restaurant owners, hotel caterers, and grocers—arrived at the sprawling Washington Market to buy their supplies. Next to arrive were the well-heeled shoppers: those who could afford the choicest cuts of meat and the freshest produce. They came in person, both men and women, or sent their cooks. By afternoon, the best goods had disappeared and prices began to fall. Now it was time for the bargain shoppers, women from middle-class and poor families, to buy their provisions. But the keenest hunters of bargains were the boardinghouse cooks, the last customers of the day, who filled their baskets with leathery steaks and slightly rancid butter.

Descriptive accounts of the New York markets present scenes of great kinetic energy. Here is one especially vivid passage from *Scribner's Monthly*:

> Choose a Saturday morning for a promenade in Washington Market, and you shall see a sight that will speed the blood in your veins,—matchless enterprise, inexhaustible spirit and multitudinous varieties of character . . . You cannot see an idle trader. The poulterer fills in his spare moments in plucking his birds, and saluting the buyers; and while the butcher is cracking a joint for one purchaser he is loudly

canvassing another from his small stand, which is completely walled in with meats. All the while there arises a din of clashing sounds which never loses pitch. Yonder there is a long counter, and standing behind it in a row are about twenty men in blue blouses, opening oysters. Their movements are like clock-work. Before each is a basket of oysters; one is picked out, a knife flashes, the shell yawns, and the delicate morsel is committed to a tin pail in two or three seconds.[12]

Artists were also drawn to the markets. Their challenge was to capture the ceaseless activity of the market in a single, unmoving image. One particularly successful illustration depicts the arrival of fresh Georgia watermelons at the Fulton Market. In this scene, a good cross-section of New York has swarmed the melon stand: barefoot street children, tramps, working men of color, housewives in bonnets, a mustachioed gentleman in a silk top hat. As the image makes clear, the markets were democratic in character, serving the broadest range of New Yorkers from Fifth Avenue tycoons to downtown street urchins.

The watermelon stand at the Fulton Street market, 1875.

The Essex Market on Grand Street, where Mrs. Glockner did her shopping, was a three-story brick building that ran the entire length of one city block. In design, it resembled a medieval fortress with massive square towers at each corner. Like other market buildings, it served more than one purpose. Food sellers occupied the ground floor, while the upper floors were home to a courthouse, a police station, a jail, a dispensary, and, in later years, a makeshift grammar school.

The Essex Market housed twenty vegetable and poultry stalls, eight butter and cheese stalls, six fish stalls, twenty-four butcher stalls, two stalls for smoked meat, two for coffee and cake, and one for tripe. In all likelihood, this is where Mrs. Glockner bought her veal bones, pig's knuckles, cabbage, salsify (a root vegetable much loved by the Germans), plums, and apples. It's also where she shopped for fish.

Predictably enough, the biggest fish-eaters in pre-modern Germany lived in coastal areas along the Baltic and the North Sea. Here, fishing boats trawled for cod, salmon, whitefish, flounder, among other forms of marine life. Their most prolific catch, however, was the diminutive herring. In its fresh form, this small, silvery fish (cousin to the sardine), figured prominently in the local diet. Preserved herring, meanwhile, became an important trading commodity. Cured in brine and packed into barrels, it traveled inland and established itself in the German kitchen. In the nineteenth century, immigrants brought their taste for herring to America, where it was never too popular among native-born citizens. Still, every winter, schoonerloads of herring arrived at the wharves along the East River and were sold in the public markets, both fresh and salted. Germans, along with the Irish, British, and Scots, were the main customers. The herring found a more welcoming home in a new kind of American food shop that began to appear on the Lower East Side sometime in the 1860s. The Germans called them delicatessens.

The delicatessen shopper could choose among herring dressed in sour cream and mayonnaise, pickled herring, herring fried in butter, smoked herring, and rolled herring stuffed with pickles. There was some version of a herring salad, a fascinating composition of flavors, textures, and colors. The following is a typical example:

HERRING SALAD

A very popular German salad is made in this manner: Soak a
dozen pickled Holland herring overnight, drain, remove the
skin and bones, and chop fine. Add a pint of cooked potatoes, half a pint of cooked beets, half a pint of raw apples,
and six hard-boiled eggs chopped in a similar manner, and a
gill each of minced onions and capers. Use French dressing.
Mix well together. Fill little dishes with the mixture, and trim
the tops with parsley, slices of boiled eggs, beets, etc.[13]

The building at 97 Orchard Street stands atop a natural elevation that
protects it from flooding, a problem that afflicted most other sections of
the Lower East Side. Thanks to that subtle rise, the building's rooftop
offered sweeping views of the surrounding neighborhood. Directly to
the east lay a tight grid of squat row houses. Here and there, one of the
newer tenements poked up awkwardly, a brick giant among dwarves. In
the courtyards formed by the grid, the square within each city block,
were additional structures, "rear tenements," as they were known, which
provided New Yorkers with some of the worst housing in the city. Closer
to the river, the rear tenements were replaced by factories (most were for
furniture), and past them, the shipyards. Beyond lay the wharves, visible
only as a thicket of ship's masts. Facing north, the grid opened slightly,
the blocks were longer and the avenues wider. The buildings were newer
and taller. Tompkins Square (Germans called it the *Weisse Garten*—the
"white garden") was among the few open spaces in the city grid. Nearing
the river, the landscape turned more industrial, the tenements replaced
by lumberyards, slaughterhouses, and breweries. To the south, toward the
narrow tip of Manhattan, lay the Five Points, a maze of skinny passage-
ways and tottering wooden houses. Just beyond it rose the domed cupola
of City Hall. To the west of Orchard Street stretched an unbroken string
of saloons, restaurants, theaters, and beer halls, some large enough to

accommodate a crowd of three thousand. This was the Bowery, New York's main entertainment district. Beyond it, Broadway, the city's widest street, sliced the island neatly down the middle.

The view from 97 Orchard embraced roughly four city wards, a geographic designation dating back to 1686, when New York's British governor divided Lower Manhattan into six political districts, each one responsible for electing an alderman to sit on the Common Council, the city's main governing body. As the city expanded northward, new wards were created, so by 1860 it had twenty-two. From the roof of 97 Orchard, the view encompassed the tenth ward (home to the Bowery), the seventeenth ward surrounding Tompkins Square, and the eleventh and thirteenth wards covering the industrial blocks along the river. Those same four wards made up *Kleindeutschland*, "Little Germany," the focus of our present story and the center of German life in New York.

The residents of *Kleindeutschland* were largely urban people. They had emigrated from cities in Germany and knew how to manage in one. (Immigrants from the German countryside generally passed through New York on their way to Missouri, Illinois, or Wisconsin, wide-open states where land was cheap and they could start farms.) New York Germans, by contrast, earned their living as merchants or tradespeople. Many were tailors, like Mr. Glockner, but they were also bakers, brewers, printers, and carpenters. Despite their shared roots, however, the residents of "Dutch-town," as it was sometimes called, were divided into small enclaves, a pattern that mirrored the cultural landscape of nineteenth-century Germany.

Maps of central Europe in the mid-nineteenth century show *Der Deutsche Bund*, "the German League," a confederation of thirty-nine small and large states. The people who made up that sprawling political body, however, were bound together in much smaller groups. Nineteenth-century Germans identified themselves as Bavarians or Hessians or Saxons. Their loyalties were regional, cemented by cultural forces like religion and language. Depending largely on where he lived, a German could be Catholic or Jewish or Lutheran or Calvinist. Germans spoke a variety of local dialects that were often unintelligible to outsiders. And each region had developed its own food traditions that the immigrants carried with them to New York.

Very broadly speaking, the culinary breakdown looked something like this: Germans from southern states like Swabia, Baden, and Bavaria depended on dumplings and noodles, a class of foods which the Germans called *Mehlspeisen* (roughly, "flour foods"), as their main source of calories. Northerners, meanwhile, relied more on potatoes, beans, and pulses like split peas and lentils. Where northerners tended to use pork fat as a cooking medium, southerners used butter. Where northerners consumed large amounts of saltwater fish, southerners ate freshwater species like pike and carp. Though Germany was a nation of sausage-eaters, every region, and many cities, produced its own local version. So, Bavarians had *weisswurst* (white sausage), a specialty of Munich, while Swabians had *blutwurst* (blood sausage) and Saxons had *rotwurst* (red sausage). The residents of Frankfurt, a city in Hesse, consumed a local sausage called *Frankfurter wurst*, the ancestor of the American hot dog. Turning to baked goods, Berlin was the city of jelly doughnuts, while Dresden produced stollen, and Nuremburg made gingerbread. And finally, the liquid portion of the meal. While beer was the national beverage, Germans also enjoyed cider, the regional favorite in Hesse, while Badeners favored wine and northerners preferred a local version of schnapps.

As they settled on the Lower East Side, Germans tended to form village-like clusters, a settlement pattern repeated again and again with successive immigrant groups. It was a precarious life, especially at first, so Germans from the same town or city banded together to form *landsmanschaften*, clubs that offered a crude but important form of life insurance. To join, the immigrant paid an initiation fee of two or three dollars, then monthly fees of a quarter or less. In return, members were invited to picnics and dances, but more important, the pooled money went to help members in distress, people who were sick or who couldn't work for one reason or another. But the *landsmanschaften's* true raison d'être was death. When a member died, the club paid for the burial—it also supplied the burial plot—and ensured a good turnout at the funeral.

Beginning in the 1850s, the Lower East Side saw a steady flow of outside visitors, among them city officials and social reformers who came to investigate tenement living conditions. Journalists flocked to the tenements in search of human-interest stories, which they found in great supply. Each of these groups set down their observations, leaving us with a large body of descriptive writing. A number of themes snake through this literature. A few of the most persistent are overcrowding in the tenements, the absence of sunlight, and the absence of fresh air, the three evils which outsiders identified as the crux of "the tenement problem." (Visitors were much less interested in the low wages and high rents that made crowding necessary.) Closely related to evil number three were the smells of the tenement, a topic that captivated uptown visitors, who prowled the East Side wards with handkerchiefs held before their noses. The following account, taken from an 1865 article in the *New York Times*, describes an interview with an East Side woman who lived in Fisher's Alley, a particularly fragrant strip in the old fourth ward:

> *We were greeted courteously by an old woman with a short garment and a pipe not much longer, and by her we were entertained with a vivid description of life in Fisher's alley. Fights, rows, scrambles for supremacy, sickness, death, much misery, but, on the whole, not so bad as it might be. Dirt in every shape, filth of every name, smells in every degree, from the faintest suggestion of fat-boiling, through the intermediate gradings of close, heated rooms, unswept floors, perspiratory and unwashed babies, unchanged beds, damp walls, and decayed matter, to the full-blown stench which arose from the liquid ooze from the privy—these combined failed to impress the speaker or, indeed, any of the slightly-clad women who joined us in the passage, as anything to feel annoyed about, and we left her with the conviction that, however wretched and offensive she was, she had at least the consolation of not knowing it.* [14]

The gulf between tenement dwellers and their uptown observers was so wide that the *Times*'s reporter felt perfectly free to share his disgust for the courteous old woman and her pungent surroundings, confident that his readers would feel the same.

Reporters generally gravitated to the worst buildings in the poorest sections, but even in a well-kept tenement the air was thick with competing odors. Especially in winter, when doors and windows were closed to shut out the cold, the tenement became a kind of hothouse in which smells bloomed, instead of flowers. In the German wards, however, one especially potent smell overwhelmed the rest: the sulfury, penetrating tang of sauerkraut.

In the patchwork that made up *Kleindeutschland*, sauerkraut was everywhere. It cut across ethnic boundaries and economic ones, too, consumed by rich and poor alike. Between late October and early December, tenement housewives (and saloon keepers as well) turned their energies to sauerkraut-making, producing enough in those few weeks to last through most of the year. In a pre-Cuisinart world, the chopping of that much cabbage was a daunting project, so women enlisted the help of an itinerant tradesman known as a *krauthobler* or "cabbage-shaver." With a tool designed specifically for the task—it worked like a French mandolin, the blades set into a wooden board—the *krauthobler* went door to door, literally shaving cabbages into thread-like strands. The cost was a penny a head.

Once the cabbage was shaved, the housewife took over. She scoured an empty liquor or vinegar barrel and lined it with whole cabbage leaves. Next came the shredded cabbage, which she salted and pounded, layer by layer, until the barrel was nearly full. Now she covered the cabbage with a cloth, then a piece of wood cut to the size of the opening, weighing it down with a stone. Left on its own, the salted cabbage began to weep, creating its own pickling brine. Once a week, the housewife tended to her barrel, rinsing the cloth to prevent contamination and skimming the brine.

Sauerkraut-making in the tenements was a harvest ritual, a celebration of the autumn bounty. Like all seasonal rites, it marked the passage of time. Its power came through repetition. The scrubbing of the barrel, the arrival of the cabbage-shaver, the salting and pounding, were all steps in a familiar routine that the immigrant housewife carried with her from Germany. Seasonal food traditions, like sauerkraut-making, supplied an uprooted community with a sense of order. At Christmas, the Germans baked squares of *lebkuchen*, or honey cake; loaves of stollen, a sweetbread studded with raisins, and trays of *pfeffernusse*, peppery spice cookies coated in sugar syrup. In spring, for just a few weeks, German saloons served up mugs of dark bock beer. Summer in *Kleindeutschland* arrived on Pentecost Sunday, which the Germans marked with an all-day picnic. Each of these food-based rites, carried over from Germany, was reenacted in a completely new context by the immigrants who settled in New York and other cities throughout the United States. Over the decades, as Germans assimilated into the wider culture, the need for the old rituals began to slip away, replaced in some cases by new American customs. But assimilation moved in the opposite direction as well. Many German food traditions were adopted by the wider culture, so baking stollen became a Christmas tradition in non-German families along with decorating the Christmas tree, another German contribution to American home life.

If fall was the season for sauerkraut-making, the payoff came in the first days of winter, when the cabbage was fully ripe and ready to be eaten. It was a moment the Germans looked forward to expectantly and enjoyed completely: "The look of pleasure on the bibulous German as he steps out of his favorite lager-beer saloon these cold days tells the passer-by as plainly as do the words that hang outside the door that the day of sauerkraut lunch is here."[15] This happy vignette is taken from a Philadelphia newspaper, another city with a large German community, but could just as easily describe the saloon-goers of Chicago, Milwaukee, or New York.

Alongside the *krauthobler*, a figure who had vanished from New York by the close of the Civil War, the German appetite for pickled cabbage also supported sauerkraut importers, local cabbage farmers, and eventually sauerkraut manufacturers, including Henry J. Heinz, who opened a sauerkraut factory on Long Island in the 1890s. At the height of the busy season, his factory processed a hundred tons of cabbage a day. On the streets, the most visible face of this trade was the "sauerkraut man," actually a roving peddler who sold cheap meals to hungry East Siders. Here he is in a 1902 article from the *New York Evening Post*:

> *The regular and popular visitor to the German inns and taverns of the East Side is the sauerkraut man. He brings his calling with him*

from the Old Country, and finds a more profitable field in New York
than in Berlin or Hamburg. His equipment is quite curious. He
wears a blue or white apron running from his neck nearly to the an-
kles, and from his shoulders is suspended a circular metal box which
goes half way around his waist. It has three large compartments, two
of which are surrounded by hot water. In one are well-cooked Frank-
furter sausages, and in the other thoroughly boiled sauerkraut. In the
third compartment is potato salad. He carries in his hand a basket in
which are small plates and steel forks. One sausage and a generous
spoonful of sauerkraut and potato salad cost 5 cents. All three articles
are of good quality, well cooked and seasoned.[16]

The sauerkraut man worked at night, his shift starting at the close of the
normal workday, when customers poured into the saloons for an hour
or two of relaxation. Hauling his pewter box (it could hold up to fifty
sausages, seven pounds of sauerkraut, and seven of potato salad), the
peddler made his rounds stopping at bars, bowling alleys, and meeting
halls, wherever hungry Germans gathered.

To round out our look into German sauerkraut traditions, here is a
recipe for a simple sauerkraut dish adapted from Henrietta Davidis.

BOILED SAUERKRAUT

Bring to a boil one cup water and one cup white wine. Add
the sauerkraut, roughly 3 cups, a few peppercorns and a
little salt. Simmer until tender. Shortly before serving, pour
off the broth and stir in a few tablespoons butter. Serve as a
side dish alongside mashed potatoes.

Nineteenth-century New York was a city of hand-painted signs, many of
them wordless. Butchers, for instance, displayed a painted black bull (or

sometimes a red cow) over their stalls in the market. Out on the street, passersby could identify a blacksmith's shop by the image of a painted horse suspended over the doorway. Even more straightforward, New York restaurants often nailed a real tortoise shell to the doorpost: their way of announcing that terrapin was on the menu. In the city's German wards, a few signs were especially common. Two yellow boots, one larger for a man, the woman's boot smaller, was the image displayed by German shoemakers. German beer halls hung pictures of King Gambrinus, the Dionysus of beer. In some of the flashier examples, the mythic king was "presented life-size, bearded and crowned and holding in one hand a stupendous beaker of the national beverage, the froth of which bulges from the rim like a prize cauliflower."[17] The description comes from Charles Dawson Shanley, a nineteenth-century poet and journalist who wrote a series of very informative articles on New York street life. On his rambles through *Kleindeutschland*, Shanley encountered another frequently displayed shop sign, this one rather modest. It was a "dingy little signboard with a sheaf of wheat painted on it"—the image adopted by German bakers.

Just as they lived together in clusters, immigrants tended to work together in the same trades. Many, as it happens, were food-related. Where the Irish were big in the fish and oyster business, Germans worked as dairymen, grocers, and butchers. Immigrant food purveyors sold to their own communities, but also played a role in feeding the larger city. Through the first half of the nineteenth century, most of the city's bakers were Scottish and Irish, but that began to change in the 1850s, as Germans flowed into New York. By the end of the decade, responsibility for baking the city's bread had passed into German hands.

The typical German bakery, housed in a tenement cellar, was a low-ceilinged room with a dirt floor and no running water. The "boss baker" often lived upstairs with his family and a handful of employees who shared the apartment as boarders. Many times, though, employees slept in the cellar next to the ovens, a sack of flour for their bed. Some slept in the dough vats. Economic survival for the small-time baker depended on every member of the family. The children worked as apprentices, while the baker's wife was in charge of the boarders, for whom she cooked and did laundry. On the most densely populated blocks in the German wards, a cellar bakery was found in every third or fourth building.

Prior to the widespread use of steam power beginning in 1882, industry in New York ran on muscle power, most of it supplied by immigrants. In a city of shipbuilders, ironworkers, and stonecutters, the baker's life was especially harsh. His shift started late in the afternoon and lasted until early morning, which meant a fourteen-hour workday or sometimes more. At the end of the long, hot night (temperatures in the bakery could easily reach one hundred degrees), the bakers hauled their goods up to the street and loaded up the delivery wagons. Now, finally, it was time to rest, just as the sun was coming up over the East River. Faces caked with flour, the bakers slept while the rest of the city went about its business. It was a topsy-turvy existence and a lonely one, too. For all his sweaty work, the journeyman baker earned between eight and eighteen dollars a week, hardly enough to support a family. The consequences were plain. More than any other tradesmen, many New York bakers were consigned to a life of bachelorhood.

Before the appearance of national brands like Pepperidge Farm and Arnold, each city had its own local bakeries and bread-making traditions. The kind of bread produced in New York was surprisingly similar to Wonder Bread, squishy and gummy-textured. Known as the New York split loaf, it was no more substantial than "slightly compressed white smoke" in the words of one critic, and just as tasteless. German-made loaves of rye and pumpernickel fell at the other end of the baked goods spectrum. They were made from whole grains, with a dense, chewy texture and a sour, mildly nutty flavor. When sliced, they made a sturdy platform for the open-faced sandwiches that Germans loved to snack on. When it came to New Yorkers and bread, a "Goldilocks syndrome" seemed to prevail. If the New York split loaf was too puffy and bland, German-style breads were too coarse and heavy for the native-born, with their less vigorous digestive tracts. The only reason to eat them was the price, since ounce for ounce they were cheaper than white bread. A brittle-crusted French baguette was much closer to the nineteenth-century ideal of what bread should be.

A footnote to the German bread story centers around a New York immigrant named Louis Fleischmann, born in Vienna in 1835. His early history had nothing to do with bread or baking. Rather, Fleischmann was a soldier, an officer in the Austrian army. In the 1860s, his two brothers, Max and Charles, emigrated to Missouri, where they set up a business producing the kind of compressed yeast used by Viennese bakers, a product unknown in America. In 1874, Louis decided to follow them. In the centenary year of 1876, Louis and his brothers set up a "model Vienna bakery" at the great Centennial Exhibition in Philadelphia. A smashing success, its main product was something called "Vienna bread." Buttery and delicate, with a glossy brown crust, it was the perfect texture for dunking in coffee. Riding on the success of the model bakery, Louis Fleischmann opened a similar establishment on Tenth Street and Broadway in New York.

The Vienna Bakery arrived on the gastronomic scene like a visiting dignitary. Alongside the actual bakery, Fleischmann opened an elegant café that quickly became a favorite dining spot among German intellectuals and opera stars. It was also popular with New York society women, who flocked to the bakery after a strenuous morning of shopping on the Ladies Mile, the strip of department stores that once ran along Lower Broadway. Of all the dishes on the menu, Vienna bread was the star attraction. When Teddy Roosevelt was police commissioner of New York in the 1890s, he used to walk uptown from his office on Mulberry Street and stop at the bakery for a lunch of Vienna bread and milk. From Fleischmann's bakery, Vienna bread spread to German bake shops around the city, but the stores most likely to carry it were on the Lower East Side. An 1877 article on the Vienna-bread phenomenon opens with the following observations:

> One remarkable result of the Centennial exhibition is the striking and admirable fact that Vienna bread is now to be bought all over New York. Indeed, we are quite sure that the genuine article is now more easily procurable in this city than in the Austrian capital. You will find it in the Bowery, and in the streets crossing that elegant

avenue; nay, you shall not enter a little baker's shop in Mackerelville
without finding at least Vienna rolls upon the counter.[18]

When Louis Fleischmann died in 1904, the Vienna Bakery had already lost its glamour, though it remained in business for several decades. The craze for Vienna bread was also starting to fade. The precise date is hard to pinpoint, but sometime after World War I, when Germans and their food fell out of favor, it began its final descent into obscurity. Even so, Fleischmann's legacy continues, visible on every packet of Fleischmann's Instant Yeast, the brand most used by American bakers for over a century.

The greatest contribution made by German bakers to the American kitchen came in the form of yeast-based cakes, which began to appear in East Side bakeries during the second half of the nineteenth century. Though all were made from the same basic dough, they came in an assortment of shapes and with a variety of toppings and fillings. There were round cakes crowned with apple slices, ring-shaped cakes filled with chopped nuts or poppy seeds, pretzel-shaped cakes, and cakes that were rolled up like snails then brushed with butter and sprinkled with cinnamon, sugar, and currants. The allure of these buttery confections quickly leapfrogged beyond *Kleindeutschland* into the wider city. The Germans called them *kuchen*, but we know them as coffee cake.

In the 1870s, the *New York Times* ran a food-related column on their women's page, called "The Household." Most columns opened with a round-up of what New Yorkers could expect to find at the market that week, which foods were in good supply, which were scarce, and current prices. The market news was followed by a selection of recipes and household tips covering a broad range of very practical topics, like how to make glue or how to stop one's shoes from squeaking. The column ended with questions and requests from readers, including this one, which ran in 1876: "I would like a receipt for pumpkin pie and German coffee-bread or coffee-cake like you get in the bakeries in New York and which cannot be found in the country.—JEWEL"[19] Unfortunately, it seems that Jewel never got a response, but over the next few decades, recipes for German coffee cake began showing up in American newspapers, magazines, and cookbooks. The following recipe for a kind of circular coffee cake called a *kranzkuchen* is courtesy of a German-American housewife who shared her kuchen-making technique with a New York reporter. It appeared in an 1897 feature under the headline "Toothsome German Dishes, Lessons To Be Learned From The People Who Eat Five Meals A Day."

KRANZKUCHEN

Take two pounds of flour, a pint and a half of milk, three eggs, a quarter of a pound of butter. Set a sponge with one pint of milk warmed, flour to make a stiff batter, and one cake of compressed yeast. When it has risen sufficiently, add the other ingredients, the butter being worked into the flour; then knead well. The cake should be rolled, or better, pressed out with the fingers very thin for baking. . . . The dough is . . . brushed over with melted butter, and upon the thin cake, sugar, cinnamon, chopped almonds, currants, and raisins are laid. The whole is rolled as a jelly cake, and then formed into a ring, Kranz, or double ring, pretzel, as desired, and also baked in a moderate oven. When this is done, a thin frosting of white of egg and sugar is spread over it, and the result is a very delicious cake, which is eaten with an excellent cup of coffee.[20]

Among the Germans' most far-reaching gifts to American food ways wasn't a food at all. . . . Beginning with the earliest settlements in Dutch New Amsterdam, our European ancestors displayed a keen fascination with the making and drinking of alcohol. There were practical reasons for this taste: drinking water during that era was often polluted. A strong taste for beer among seventeenth-century New Yorkers gave way in the following century to the widespread consumption of locally produced rum, the same throat-scorching drink that played such an important role in the colonial slave trade. (Throughout the eighteenth century, rum was the single most important American export, much of it shipped to Africa and traded for living cargo.) In the years following the revolution, as the American rum industry began to falter, farmers in western states like Kentucky, Tennessee, Indiana, and Illinois turned their crops into liquor. Corn-based bourbon and grain-based whiskey soon eclipsed rum to become the new national drinks. Through most of this period, Americans continued to gulp down homemade forms of alcohol, including cider, apple jack, and dandelion wine. The one drink they did not have was lager, the crisp German-style beer so familiar to us today.

Before 1840, all beer produced in the United States was English-style ale, full-bodied and slightly fruity-tasting, with a deep caramel color and a high alcohol content. Following British brewing methods, it was made with a type of yeast that floats on the surface of the brew and ferments rather quickly at relatively high temperatures. German immigrant brewers brought to the United States a separate brewing tradition that was based on a different strain of yeast, one that sinks in the vat. Known as bottom yeast, it ferments more slowly and at much lower temperatures, producing a beer that is dryer, paler, and more refreshing than ale.

The first German breweries in New York were small operations employing five or six men. German brewers followed the same basic plan as immigrant bakers: under the watchful eye of a skilled brew master, the brewery workers put in sixteen-hour days of hard labor. In return, they received a small salary (between six and twelve dollars a month) plus room and board, along with all the beer they could drink. Since it took roughly one thousand dollars to open a brewery, the lager entrepreneur—unlike other immigrant businessmen—was a person of means. The great majority were established beer manufacturers who brought to America a lifetime of brewing experience, including their own closely guarded brewing formula. Two of the first to get started in New York were a pair of German brothers, Max and Frederick Schaefer, who opened their Manhattan brewery in 1842. At the time, most of the city's lager drinkers came from within the German community, but that was soon to change:

When lager was first introduced to New York by the Schaefers, they kept a tavern on Seventh Avenue, between Sixteenth and Seventeenth streets. Americans, hearing the praises of the new beverage and seeing their Teutonic friends roll their eyes and smack their lips in ecstatic contemplation and enjoyment of it, used to make bold essays at its consumption, with the almost universal result of being intensely disgusted by its novel bitter taste. With many contemptuous ejaculations, wry faces, and much sputtering and rinsing of their mouths with the familiar whisky, they would revile and condemn the Deutscher's delight. The first beer saloon downtown was started on Broadway, just a little below Canal street, where similar experiments and disappointments were long the order of the day. Soon after, lager beer saloons appeared with almost magical rapidity all over the City. New York seemed to have broken out with a rash of them scarcely more than five years later. It was not until about 1855, however, that any great number of Americans took kindly to the German drink. Gradually they began to like the stuff.[21]

The Schaefers represent only one of the German beer-making dynasties to emerge in nineteenth-century America. The complete list contains some very familiar names, including Frederick Miller, Adolphus Busch, Captain Frederick Pabst, and Joseph Schlitz.

For uptown New Yorkers, a cool glass of lager was the ideal warm-weather drink. For residents of *Kleindeutschland*, it was a daily staple, a fact which non-Germans marveled over: "They drink it in the morning, at noon, in the evening and late at night, during their labors and their rest, alone and with friends . . . They take lager as we do oxygen into our lungs—appearing to live and thrive on it."[22] Just about every block in the German wards had at least one beer saloon, establishments where men like Mr. Glockner went to read the daily papers, play cards, talk politics, and conduct their business. The East Side saloons were the working man's version of a private club:

*A German must have time for his libations. He cannot march up
to the bar, pour out a drink, dash it down without the possibility
of tasting it, toss the money over the counter, and rush out like an
ignited sky-rocket, as the majority of Americans do. Tables, chairs,
newspapers, cigars or pipes, and friends are not merely comfortable
additions, but actual essentials to his enjoyment. Instead of a quarter
of a minute he wants at least a quarter of an hour for the proper
enjoyment of a drink. Conversation is another essential. However
taciturn the German may appear among others, let him sit down at
one of these tables and get his glass of lager beer and a listening friend,
and if anyone desires to know how much talk a human tongue can
reel off in any given period, then is the time to listen.*[23]

Americans consumed their alcohol with a rebellious drink-to-get-drunk
attitude. One result of that was the close association between drinking
and brawling in American life. The Germans, by contrast, reveled in the
communal spirit that developed after a glass or two of lager. They drank
methodically, pacing themselves like marathoners to wring out every possible ounce of pleasure. Though Americans adopted beer as their national
drink, they never fully acquired the Germans' flair for savoring it.

For German home consumption, parents would send one of their
kids down to the local saloon with a tin pitcher or pail—East Siders
called them "growlers"—which the barkeep would fill for around fifteen
cents. The sight of young East Side kids shuffling home with growlers
full of beer was commonplace enough to catch the attention of Jacob
Riis, New York's best-known social reformer. In his now-classic *How the
Other Half Lives*, Riis offers a possibly apocryphal story about one East
Side boy, who spent his Saturday ferrying growlers to his father's workplace. By evening, the kid was so drunk he disappeared into a cellar to
"sleep off the effects of his own share in the rioting." On Monday morning, after a weekend of desperate searching, the boy was discovered by his
parents, dead and half-eaten by rats.[24]

In contrast to their American neighbors, the Germans saw beer as a
family drink. On Sunday afternoon, entire immigrant families (babies

included) celebrated their one day of leisure with a trip to the cavernous beer halls that lined the Bowery. The largest and best known was the
Atlantic Gardens—a somewhat misleading name, since it wasn't a garden
at all, but a long, barrel-vaulted room large enough to hold a blimp, or
maybe two. It was a highly functional space, designed to house as many
people as possible. From the floor to the top of the ceiling, every interior surface was adorned with an intricate pattern of swirling plaster
medallions and curlicue borders. The hangar-like proportions of the hall,
combined with the fancy plasterwork, gave it the feel of a gilded shed. A
raised gallery that projected into the room provided a stage for musicians.
During the day, sunlight streamed in the hall from skylights at either end
of the building. At night, it glowed with the light of three gas-burning
chandeliers, each one of them six feet in diameter.

On a pleasant Sunday afternoon, when the room was full to capacity,
the level of activity inside the Atlantic Gardens must have been dizzying.
As an all-female band played from the gallery, a crowd of three thousand men, women, and children were drinking, talking, and laughing.
The youngest family members, babies who were too young to sit at the

A typical Sunday in a German beer garden, 1872, this one located on the Bowery.

table, were plunked on the floor by their mothers' feet, where they presented a tripping hazard to the hurrying waiters, their trays loaded with beer mugs.

Chroniclers of nineteenth-century New York were drawn to the beer halls for the vivid subject matter they provided. They marveled at the vastness of the rooms, the quantity of beer consumed in a single business day, and the quality of the house musicians. The one subject no one seemed to care too much about was the food. The most we can say is that it was hearty and simple. Brown bread seasoned with caraway, plates of Swiss and Limburger cheese, sliced ham, pickles, and salted pretzels were all standard beer-hall fare. Some patrons brought their own picnic-style snacks of bread and sausage or bread and cheese—a practice welcomed by the beer-hall management so long as they paid for the drinks.

A more elaborate meal awaited diners at the German "lunch rooms" that once thrived in New York. Down near the tip of the island, in the heart of the financial district, the lunch rooms served an all-male clientele of shipping agents, bankers, lawyers, and insurance brokers. Mixed in among the businessmen were a scattering of journalists and engravers from nearby Printing House Square, where all the city newspapers had their offices. In fact, one German lunch room, the Rathskeller, was housed in the basement of the *Staats Zeitung*, New York's largest German-language newspaper. There was also the Postkeller at the corner of Broadway and Barclay, Hollander's at the intersection of Broadway and Chambers, and Dietz on North William Street. At these eateries, customers could dine on a bowl of soup, a cut from a joint of meat, vegetables, salad, and a glass of beer, all for 35 cents. Ermich's, at the corner of Nassau and John streets, was a crowded basement room with large communal tables where diners could start their meal with a bowl of smoked-sausage-and-lentil soup. If they wanted bread for dunking, they cut off a hunk from a shared loaf at the center of the table. (Etiquette at Ermich's demanded that each diner wipe his knife across the top of the bread before cutting his slice.) Entrées included fish balls smothered in red cabbage, Vienna sausage with "half-half" (a side dish that was half mashed potatoes and half sauerkraut), or a "fricatelleu [stew] of minced meat surrounded by a brownèd crust composed of equal parts of flour and potato." There was also schnitzel (fried veal cutlet), sliced tongue with raisin sauce, and a dish called "Hamburger steak," a form of ground beef "redeemed from its original toughness by being mashed into mincemeat and then formed into a conglomerated mass."[25] This not-too-appetizing description is among the earliest reference to a future American staple, the hamburger, seen here at the very start of its culinary journey.

The German restaurants that proliferated in nineteenth-century New York appealed to both immigrant diners and native-born citizens. In fact, of all the city's foreign restaurants, French and Italian included, Americans showed the greatest admiration for those owned by Germans. Their fondness reflected how they perceived the immigrant himself. Until World War I, when global politics recast Germans as "enemy aliens," Americans considered Germans the model immigrants—industrious, intelligent, highly cultivated, and impeccable in their personal hygiene. Temperamentally, the German was jovial and attentive to his guests, qualities that made him an ideal restaurant host. And unlike Americans, constrained by their puritan discomfort with bodily pleasures, Germans knew how to relish their food. "Of all the foreign elements in town," one New Yorker remarked, "none delight more in good eating and drinking than the Germans."[26] In the 1870s and 1880s, the hundreds of German restaurants scattered through the city became popular gathering spots for New York businessmen, professionals, and "clubmen," who spent their leisure hours in the city's many gentlemen's clubs. The most celebrated German establishment was Sieghortner's, located in the old Astor mansion on Lafayette Place. Guidebooks to New York listed Sieghortner's among the city's premier dining spots, ranking it second only to Delmonico's.

Around the corner, at Bleecker and Broadway, a very different crowd was assembled at a German restaurant called Pfaff's. Named for its owner, Charles Pfaff, it served as the unofficial headquarters for the city's "Bohemian" set, making it one of the first "ethnic" eateries to attract socially prominent New Yorkers. Of course, Swiss-owned Delmonico was "ethnic" too, but in the most refined and elegant way possible. Pfaff's, by contrast, was a dive. It opened early in the 1850s in a dingy and ill-ventilated space that was literally under the Broadway sidewalk. But that was part of its charm. The subterranean location gave it a hidden-in-plain-sight kind of allure captured by Walt Whitman, an honorary Bohemian and steady Pfaff's customer:

> *The vault at Pfaff's where the drinkers and laughers*
> *meet to eat and carouse,*
> *While on the walk immediately overhead pass the*
> *myriad feet of Broadway.*[27]

The part Whitman leaves out is that customers could actually look up and see the shadowy forms of passersby, visible through glass bull's-eyes that had been set into the pavement.

The Bohemians who gathered at Pfaff's were the beatniks of their time. Self-proclaimed rebels, they laughed at bourgeois respectability and flaunted social convention, sexual and otherwise. Their ringleader was Henry Clapp, a newspaperman who returned from a trip to Paris fired up by Henry Murger's 1851 novel, *Scènes de la vie de Bohème*, which is where the term "Bohemian" originates. His followers included Ada Clare (a writer and actress), Edward Wilkins (drama critic for *The Herald*), George Arnold (essayist and poet), Artemus Ward (a humorist), and the poet Walt Whitman, who was more a revered spectator than a full-fledged participant.

An unobtrusive but sympathetic character, Charley Pfaff, owner and host, became a minor celebrity in his own right. Contemporaries said he ran the best bar in New York, stocking it with a broad selection of the finest European wines. He was better known, however, for his imported beer, the beverage of choice among his Bohemian clientele. As you might imagine, the Bohemians liked to arrive late. Their midnight suppers consisted of oysters, steak, liver and bacon, and Welsh rarebit, the typical foods of a New York chop house. Alongside these American staples, the kitchen prepared "foreign" specialties like *pfankuchen*, Frisbee-sized German pancakes. It was a dish admired by the Ohio-born novelist William Dean Howells on his visit to Pfaff's in 1860.

German Pancakes

One heaping cup flour, ½ teaspoonful salt, 2 cupfuls milk
or water, 3 eggs. Sift flour and salt into a bowl, add the
milk and the 3 yolks, mix it into a smooth batter; beat the
3 whites to a stiff froth, add gradually the batter to the
beaten white while stirring constantly. Place a medium-sized
frying pan over the fire, with ½ tablespoonful butter or lard;
as soon as hot, pour in sufficient of the mixture to cover
the bottom of the pan, shake the pan to and fro and bake
till light brown on the underside, turn over and bake the
other; slip the pancake onto a hot plate, bake the remaining batter the same way, and serve at once. This will make 4
pancakes.[28]

The revelries under the sidewalk lasted until 1861, when the start of the
Civil War effectively broke up the Bohemian circle. Though the sparkle
was gone, Pfaff's remained open at the same address (653 Broadway) for
another fourteen years, then moved uptown to West 24th Street, following the city's shifting center of gravity. The new restaurant was a money-
losing proposition and closed for good in 1887, three years before the
death of its genial and once-famous owner.

When Pfaff's first opened in the 1850s, New York's main entertainment district ran along Lower Broadway in the neighborhood that eventually became SoHo. Over the next quarter-century, as the city expanded
northward, the Broadway theaters began to migrate uptown, pausing for
a while at 14th Street.

For roughly three decades, the blocks between Third Avenue and
Broadway on 14th Street provided New Yorkers with a broad range of
diversions, from opera performances at the Academy of Music to musical

comedies starring Lillian Russell at the New Fourteenth Street Theater. When the shows let out, all of those theatergoers needed somewhere to eat. Their first choice was Luchow's, one of many saloons that lined the wide corridor of 14th Street. All of these businesses were immigrant-owned and -run, serving up German, Austrian, and Hungarian cooking to a mixed crowd of fellow immigrants, native New Yorkers, tourists, and visiting artists from around the world.

One vivid portrait of Luchow's comes from a hard-drinking New York journalist, Benjamin DeCasseres, who looks back on his favorite saloon through the lens of Prohibition. Writing in 1931, he remembers his first visit to Luchow's nearly forty years earlier, when he was a young reporter just discovering New York:

> *The dark wood, the high ceiling, the ultra Teutonic waiters, the dripping bar, the mounded free lunch, the heavenly odor of pig's knuckles, sauerkraut, and Paprika Schnitzel—all of the things saturated me with an indescribable feeling of contentment.*
>
> *I anchored at the bar and discovered at once that the quality and the upkeep of the beer in the place were all that my crapulous newspaper friend had told me. As it went down—Seidel after Seidel—every atom of my body bloomed with radiant philanthropy. Dill pickles and tiny raw onions burst in my throat and sprayed my brain with a fine tickle.[29]*

DeCasseres was a fixture at the Luchow's bar, where the Pilsners "tasted like moonlight." He also spent time in the dining room, eating his way through the Luchow's menu. Here are the dishes he remembered most fondly:

Veal Schnitzel with Wild Mushrooms
Boiled Beef with Horseradish
Bratwurst with Sauerkraut
Sauerbraten with Potato Dumplings
Stewed Goose with Calves' Feet

Pan-fried Hamburger
Young Pigeon with Asparagus

Luchow's specialized in roast goose, duck, and venison, and was likewise known for its crisp potato pancakes and homemade frankfurters.

The Academy of Music closed in 1886, forced out of business by the
newly built Metropolitan Opera House. In time, the rest of the theaters
followed suit. Some closed permanently, others moved uptown to the
more fashionable entertainment district around 42nd Street. Robbed of
their customers, the saloons vanished too. Incredibly, Luchow's managed
to hang on until 1982, haunting 14th Street like a stranded visitor from
another time.

The journey from Ermich's lunch room to Pfaff's to Luchow's
traced a rough semicircle that skirted the edges of *Kleindeutschland*. All
three establishments played a major part in the transmission of German
food ways to mainstream America. The frankfurters and hamburgers
eaten in similar nineteenth-century German restaurants have become so
thoroughly assimilated that we hardly recognize them as German at all.
Within *Kleindeutschland* proper, however, immigrants like the Glockners
patronized smaller, less glamorous eateries where the crowds were exclusively German. Though scattered throughout the German wards, they
were especially thick along the Bowery, the center of downtown nightlife,
and on Avenue A, the Germans' restaurant row.

More than the Irish or Russians or Italians, the Germans saw eating
as a public activity, an occasion to leave the tenement and venture into
the larger world. The typical immigrant restaurant displayed its offerings
along the bar, which doubled as a buffet counter, the food arranged like a
Flemish still life. A visitor to one such eating place, stunned by the copi-
ous display, described how the bar was

> piled with joints and manufactured meats adapted to the strong Ger-
> man stomach;—enormous fat hams, not thoroughly boiled, for the
> German prefers his pig underdone: rounds of cold corned beef, jostled
> by cold roast legs and loins of veal; pyramids of sausages of every

known size and shape, and several cognate articles of manufactured
swine meat. . . .

There were also baskets of freshly baked pretzels, mounds of Swiss and
Limburger cheese, heaps of sliced onion, earthenware jars of caviar and
a large glass jar of pickled oysters. The attached dining room, its walls
painted with mountain scenery, was busy throughout the day and deep
into the night, a feeding ground for German tradesmen, doctors, lawyers,
and merchants, often accompanied by their wives and children. Both at
home and in public, Germans preferred to dine as a family.[30]

But public dining on a grand scale was connected with the many clubs
and societies that formed the core of German social life in nineteenth-
century New York. Known as *Vereine*, they were a carryover from the Old
Country. The *Vereine* developed in Germany during the late 1700s, part
of a new city-based culture in which merchants and tradespeople banded
together in professional and political associations, representing their
interests as the new German middle class. As German immigrants recreated the *Vereine* in New York, the clubs lost their political edge, or most
of it, and became more purely social.

Just about every New York German belonged to at least one *Verein*,
and some belonged to many of them, especially if they were reasonably
well-off and could afford the membership fees. Any shared experience or
common interest was reason to join a *Verein*. Some were organized around
place of origin, while others were based on occupation, like the German
grocers' or brewers' *Vereine*. Common-interest *Vereine* were for serious lov-
ers of poetry or music or drama or athletics, though some were based
on the flimsiest of excuses. In the 1880s, a group of German Jews living in Harlem established a *Schnorrer's* (Yiddish for "moocher") *Verein*,
which hosted an annual clambake. Among the most prominent *Vereine* in
Kleindeutschland were German singing societies like the Arion Club, which
sang President Lincoln's funeral hymn on the steps of City Hall as his
body lay in state in the Grand Rotunda.

The *Vereine* met in saloons, beer halls, and other public spaces, like the
Germania Assembly Rooms on the Bowery or the Odd Fellows' Hall on

Forsyth Street. The larger clubs had their own private headquarters, some of which are still standing. A building on St. Mark's Place in the East Village still carries the inscription *Deutsche-Amerikanische Schutzengesellschaft*, one of the neighborhood's many shooting clubs. The clubs staged musical performances, athletic demonstrations, and theatrical shows. Fond of processions, they were often seen parading through the streets of New York carrying banners or torches. During the winter holiday season, they held masked balls (the reason *Kleindeutschland* had so many costume shops) and elaborate banquets. In summer, individual clubs joined forces, hosting enormous *Volksfest* that combined all the Germans' favorite activities: eating and drinking, shooting and athletics, singing and dancing.

The *Volksfest* began with a procession as the immigrants traveled en masse to one of the downtown ferry landings. Some societies paraded in costume. The most dashing belonged to the German *Turnverein*, a club that joined progressive thinking and gymnastics in one overarching philosophy. The headquarters for the *Turnverein* was at 27–33 Orchard Street in a building that once served as a Quaker meeting house. When the paraders left the hall, heading uptown on their way to the East River, they would have marched directly in front of 97 Orchard. When they did, the Orchard Street tenants must have run to their window to watch the passing show. This one took place in the summer of 1862, a year before the building went up:

> *The procession formed a gallant and striking spectacle. The Turners, in their uniforms of white jackets and pants, with gay kerchiefs tied around their necks, and neat black Kossuth hats, the many richly embroidered banners, the numerous and well-trained music corps, the companies of happy looking cadets, and then, along the sidewalks, the accompanying throng and interminable train of buxom women, nearly all lugging along huge chubby-cheeked babies or followed by troops of roly-poly children, some of whom were just able to toddle—all these indispensable and inevitable features of the German "Fest-tag" were there in rare profusion, brightening even the August sunshine to a ruddier glow.*[31]

When the paraders arrived at the ferry landing, they were literally shipped off to the picnic ground in Hoboken, New Jersey, or Upper Manhattan, which still offered large tracts of undeveloped land.

At the largest festivals, the number of people might reach twenty or thirty thousand. The amount of food and drink required to satisfy a crowd that size must have been staggering. Equally daunting were the logistics of preparing and transporting it in an era before the mechanized kitchen and takeout containers. It is unclear exactly who supplied the food at the picnics, though some of it came from restaurant kitchens inside the various parks. It was sold from booths or stands that were set up in shady areas, usually under a tree. The kinds of food consumed are already familiar to us. What's new is the quantity: colossal mounds of herring salad, heaps of sauerkraut the size of small haystacks, and giant sheets of honey cake. But the items that seemed to dominate the picnics were sausages, potatoes, and beer. At a picnic in Brooklyn's Ridgewood Park,

> *In all directions were arrayed booths and stalls in which edibles were prepared and offered for sale. Frankfurter sausages were in great demand, while the supply of potato pancakes was something enormous. The consumption of lager beer kept the brewers' carts busy continually coming and going with kegs of beer.*[32]

The vendors at a picnic in Harlem provided "sausages of different sizes from the small one of a finger's size to the enormous wurst two yards long and two feet in circumference."[33]

American visitors to the German picnics were awed by the sight of so many people, eating, drinking, dancing, shooting their rifles, and generally celebrating. In June of 1855, the *Saengverein*, one of the German singing societies, hosted a picnic at Elm Park on Staten Island. A New York journalist who was sent to cover the event returned with the following report:

> *We never saw the like before! Such a pouring down of lager bier, such swarms of Germans, such extravagantly jolly times, we should*

not have expected to see if we planned a year's travel and tarry in
Germany as we saw yesterday in Elm Park . . . Before 1 o'clock
from twelve to fourteen thousand persons were already busily enjoy-
ing themselves, and still for a couple of hours longer they kept stream-
ing in at both entrances.[34]

Visitors were equally impressed by the spirit of orderliness that prevailed
at the picnics, despite the size of the gathering and the amount of drinking that took place. Somewhere in that frolicsome but well-mannered
mob were the Glockners, Wilhelmina sitting under a tree on the slightly
damp earth. Her infant son nestled beside her, lying on his father's out-
spread handkerchief. The two of them are waiting for the return of Mr.
Glockner, who has temporarily disappeared into the crowd in search of
grilled bratwurst.

CHAPTER TWO

The Moore Family

Potatoes——! Kindly Root, most Cordial Friend,
That Ever Nature to this Isle did send!
Potatoes; oh hard Fate! all dead and gone?
And with them thousands of our selves anon!
'Twas you, deceas'd dear friends, kept us alive,
Vain, vain are all our Hopes long to survive!

—Anonymous eighteenth-century Irish poet[1]

It was the size of a boy's fist, though not so well-formed. Its crackled skin was covered with sunken, purplish-black marks, splattered like paint. Sliced in half, the pale interior was marbled with thick, rust-colored veins. Poke it with your finger, and the spongy flesh oozed a foul-smelling liquid. . . .

This lopsided, rotting form was a blighted Irish potato, victim of the fungus-like parasite *Phytophthora infestans.* The blight that struck Ireland in the mid-nineteenth century, triggering the deadliest famine in the history of modern Europe, originated in central Mexico sometime around 1840. From there, it migrated to the United States in 1843, first detected by

New England farmers who were mystified by its lethal handiwork. The following year, it was carried to Europe with a shipment of seed potatoes, spreading through Belgium, Germany, France, and England, then leaping over the sea to Ireland in 1845. Here, it found ideal growing conditions—cool temperatures and plenty of rain. According to Irish observers, the parasite worked on its host with such efficiency that entire fields, green and healthy one day, were black and withered within the week. The blight returned in 1846, this time causing a total failure of the Irish potato crop, the national staple, leaving most of the country with virtually nothing to eat. Incredibly, it endured for nearly a decade, hibernating over the winter then blooming to life each spring until it was finally killed off by a warmer, drier weather pattern.

In 1840, the population of Ireland was over eight million, and still quickly growing. Within fifteen years, more than a quarter of that number had vanished. One and a half million Irish had died of starvation, or of famine-related diseases, such as typhus or cholera. Approximately two million were lost to emigration, some to England and Scotland but most to America, producing another demographic shift, this one in New York.

When the nineteenth century began, New York was in many ways an English city. Most of its inhabitants were of English descent; they belonged to the Church of England, drank English-style ale, and paid for it with English-named coins. The great influx of Irish immigrants between 1845 and 1860 changed the city's ethnic composition, so by 1860 it was a quarter Irish, and nearly a third Catholic. Included in that exodus were Bridget Meehan and Joseph Moore (her future husband), the focus of our present story.

Bridget came first, in 1863, and Joseph in 1865, the final year of the American Civil War. As far as we know, both traveled alone, sailing from Liverpool to New York. If they had good weather, the transatlantic voyage took each of them roughly twelve days. Most of that time was spent in the ship's steerage, a space designed for the transport of cargo, not people. A low-ceilinged deck of six feet or less, steerage was typically divided into three sections: one for single men, another for women, and the third for families. The portholes were the only source of light and

fresh air, and during bad weather even they were shut tight. The passengers in steerage, a mix of Irish, English, Scots, and Germans, slept on
bare wooden berths six feet long and eighteen inches wide. After 1849,
ships sailing from Britain were compelled by law to allot each steerage
passenger sixteen square feet of space, a good indication of the crowding
that existed before the statute was passed.

Two or more weeks at sea, confined in tight quarters, posed consider-
able health risks. Steerage passengers regularly came down with typhus,
also called "ship's fever," and many died en route. An incubator for disease, the steerage compartment was also a tinder box for the emotions
of frightened, drunk, frustrated, and bored travelers of diverse backgrounds, suddenly thrown together. The following travel advice came
from a newly landed immigrant named Mary McCarthy, writing back to
her family in Ireland:

> *Take courage and be determined and bold as the first two or three days
> will be the worst to you and mind whatever happens on board keep
> your own temper, do not speak angry or hasty. The mildest man has
> the best chance on board.*[2]

Mary also recommends that her family travel with a bottle of whiskey,
doling out an occasional glass of it to the ship's cook and the sailors, as it
could do "no harm." But life in steerage wasn't all grim. Passengers enter-
tained themselves with card games, singing, music, and dancing. In the
evenings, as an accordion player pumped out tunes, Irishwomen danced
reels in the aisles. Children clapped to the music, while men drank toasts
to the promise of their new lives in America, filling the compartment
with a fog of blue tobacco smoke.

As to the food in steerage, travelers have left us with dramatically con-
flicting reports. In the first decades of the nineteenth century, passengers
were responsible for supplying their own provisions, and for cooking
them, too. On the upper deck, in a primitive kitchen known as the steerage galley, they boiled potatoes, oatmeal, salt beef, and water for tea, their
cooking pot suspended from a hook over a hot grate. They ate and drank

from tin cups and plates, two more items they were required to provide for themselves. (A thin straw mattress, a pot, tin plate, fork, and spoon became known as the "immigrant's kit.") Another way for the shipping lines to increase their profit, the bring-your-own system was fraught with problems. The steerage galleys were much too small, and fights erupted as families vied for cooking time. A more serious drawback was that many steerage passengers were simply too poor to provide for themselves and either brought nothing or ran out of food before the ship reached its destination.

To save the immigrant passenger from starvation and other dangers at sea, statutes were passed in both the United States and Britain to improve and regulate steerage conditions. One result of that effort was that, beginning in 1848, ships were required to furnish each passenger with the following: sixty gallons of water, thirty-five pounds of flour, fifteen pounds of ship's biscuits, and ten pounds each of wheat flour, oatmeal, rice, salt pork, and peas and beans. It was just enough food to last through the transatlantic journey which, at the time, averaged thirty-five to forty days. The steerage galleys were still mobbed, however, and wholly inadequate to meet the demands placed on them. In desperation, passengers ate their food raw, mixing flour and water into a paste and gulping it down as best they could.

By the time of Bridget's journey, it had become routine for ships to provide their passengers with cooked meals. They were eaten down in steerage on collapsible tables that were no more than crude boards balanced on trestles. On some ships, tables were lowered from the ceiling into the aisles, creating an impromptu dining room that was hoisted back up when the meal was over. Three times a day, stewards descended the narrow staircase with oversize cooking vessels, and dished out the contents. For breakfast, there was porridge with molasses, or salt fish; for lunch, boiled beef and potatoes; and for dinner, bread or biscuits and tea. Even if the meat was rank and the bread moldy, to a half-starved Irish peasant the quantity and variety was extravagant, a good omen for all the good eating that lay ahead.

The biographies of Bridget Meehan and Joseph Moore are representative of the larger Irish migration on several counts. First and most important, both arrived in New York young and unmarried: Bridget was seventeen, and Joseph twenty. Where other national groups—the Germans, Italians, and Russians, for example—settled in the United States as families, the Irish migration was essentially a movement of teenagers. Though many sent money home to bring over brothers or sisters or cousins, parents were generally left behind. The Irish were also the only major immigrant group in which women outnumbered men. Amid the cultural and economic changes in post-famine Ireland, women's status declined drastically. For many, a relatively inexpensive ticket to America was the only way to improve their lot.

Very little is known about Bridget's life during her first two years alone in New York. If she were fortunate, the still "green" Miss Meehan had a cousin or some other relative already in America to unravel the mysterious workings of her adopted home. The hard-edged geometry of the American city was utterly alien to the Irish immigrant. The ceaseless motion of both men and machines tested the newcomer's very sanity. One young Irishman, writing home to his family in 1894, tried to convey the strangeness of his current home. "This country is very different from the old one," he tells them:

> The houses are of brick five to nine stories high with flat roofs on which people walk as in a garden . . . The streets here are paved with stones, and as they are filled with fast-driven vans at all hours—the streets are as bright by night as day—the din and uproar is something horrid. Add to this the elevated railways running 30 feet above the avenues, as the cross-streets are called, the trains flying after one another like furies, and thousands of factories and steamboats whistling and roaring all the time.[3]

Letters from home were a salve to the disoriented and uprooted immigrant, but the waiting time between letters brought its own torments. A young Irishwoman living in Brooklyn in the 1880s hints at the terror she experienced waiting for the letter that never seems to arrive:

My dear Mamma,
What on earth is the matter with you all, that none of ye would think
of writing to me. The fact is I am heartsick fretting. I cannot sleep
the night and if I chance to sleep I wake up with frightful dreams. To
think it's now going and gone into the third month since you wrote to
me. I feel as if I'm dead to the world.[4]

What the lonesome immigrant seemed to crave most was information about relatives and friends in Ireland. Writing to her sister back home, a young New York immigrant named Mary Brown asks for news on aunts, uncles, nephews and nieces, siblings and neighbors, mentioning each by name and wondering who among them is in good health, who is sick, and who has been married. Just before signing off, Mary asks her sister to send her a locket of hair as a tangible keepsake.

Mary Brown worked as a domestic for a family on West 13th Street. Most likely, Bridget Meehan began her life as an American wage-earner in just the same way, as a servant, a maid or pantry girl, working for a New York family that also provided her with room and board. Disparaged by native-born citizens, domestic service was a form of work open to immigrants and people of color, and by the 1850s it was dominated by Irishwomen.

The Irish found domestic jobs through other immigrants, who acted as unofficial employment agents. Working in America as maids or cooks, they spread the word that an honest, industrious relative back in Ireland was looking for a domestic position, so when she arrived, a job was waiting for her. If she had no connections, a newly arrived immigrant could find work through a commercial employment agency, or "intelligence office," as they were known, many of them located downtown, near the docks. The typical intelligence office demanded money from the immigrants they were supposed to help, charging a fee—between 50 cents and a dollar—just to register, and additional fees after that. Even shadier, some respectable-seeming offices were fronts for less wholesome activities. Young girls who stumbled into them expecting to find work with a local family were sent instead to one of the many hundreds of brothels that once flourished in New York. To protect work-hungry immigrants, in 1850 the New York State Commissioners of Emigration opened the Labor Exchange, an office that served both women and men looking for work in New York, or anywhere else in the United States. The majority of people who registered with the Exchange were unskilled workers in low-paying jobs. The men were laborers, and the women servants, mostly German and Irish. Bridget Meehan may have been among them.

The demand for immigrant servants in nineteenth-century America was insatiable. If a household was well-to-do or even middle-class, its every function was in the hands of domestic workers. Beyond cleaning, servants were responsible for laundering and ironing, for lighting lamps, fireplaces, and furnaces. They took care of the children, nursed the sick, received visitors, and cooked and served the family meals. The house-wife's job was to manage her staff, even though she may have had no hands-on experience of the tasks they performed. If a servant suddenly quit or was hurt or sick, the household was thrown into a tumult until a replacement was found. If the family cook came down with the flu, the housewife was unable to step in and fix dinner, because she had never learned how to cook.

With so much riding on their staff, the job of finding good servants was a much-discussed topic among nineteenth-century housewives. In the second half of the century, running debates on the "servant ques-tion" appeared in the women's advice columns, now and then boiling over onto the editorial page. One question of enduring interest was: "Which nationality makes the *best* servants?"

As a rule, housewives looked on their immigrant servants as partly formed and childlike beings. The word they used for it was "raw." A German or English or Swiss girl, newly landed, was raw in exactly the right way—untainted and malleable. Raw Irish maids, by contrast, were "Dirty, impudent, careless, wasteful, and for incompetence they take the premium, but what can you expect when most of them are just off the 'bogs'?"[5] The critique comes from a New York homemaker, venting her domestic frustrations in a letter to the editor. Her biting words, one isolated expression of much broader anti-Irish and anti-

Catholic feelings that had taken hold of America, placed the Irish servant in a highly peculiar position. The same women who battered their
Irish maids with insults also relied on them to keep their families clean
and fed. Scorned and ridiculed, the Irish maid was also indispensable,
a fact she came to grasp, using it to her advantage in ways small and
large. When a servant was applying for a new job, it gave her a clear
negotiating advantage over her much wealthier American mistress. It
also made her exceedingly hard to get rid of. Despite all their moaning,
American housewives, in the interest of domestic stability, were reluc-
tant to fire their Irish maids, among the lowest paid of any nationality.
Besides, the job of finding a new one was so onerous, it was ultimately
less trouble to put up with the servants they already had, despite their
glaring imperfections.

According to American householders, the incompetence of the Irish
servant reached its fullest expression in the kitchen. Her specialties:
blackened steaks, scorched coffee, gummy puddings, leaden pastries, and
broken china, only the most expensive pieces. This ceaseless griping was
in part a matter of prior experience—or, rather, the lack of it. Most
nineteenth-century Irishwomen arrived in the United States with very
limited culinary skills. If they were country people, as many were, they
knew how to cook over an open peat-fire but had never used a stove, or
even seen one. Indoor plumbing was equally alien, not to mention the
foods themselves. Beyond boiling, they had scant knowledge of culinary
technique. It was a cruel irony that "domestic cook" was one of the few
jobs Americans were willing to grant them.

Living and working among the native-born, observing their domestic
habits at very close range, the Irish servant received a crash course in the
food culture of middle-class America. At the most nuts-and-bolts level,
she was tutored in the mechanics of the American kitchen. For the Irish,
that included how to operate a coal-burning stove, a contraption most
had never seen before. More abstractly, she was introduced to American
food traditions and assumptions. She learned, for example, what to feed
a growing child (cheap cuts of beef and lamb, "mild" vegetables like peas
and carrots, home-baked brown bread), what constitutes a nourishing

breakfast (porridge, mutton chops, fish steaks), what foods should be served at a ladies' luncheon (raw oysters, bouillon, lobster, sweetbreads in pastry), and what to cook for the Thanksgiving table (more oysters, roast turkey, chicken pie, creamed onions, mince pie, pumpkin pie). When she quit her job to marry and start a family, the typical pattern among Irish servants, she brought her knowledge of American food ways into her new life, applying it, piecemeal, to her own cooking.

Joseph Moore began his working life in America as a waiter, another job that often fell to immigrants or people of color. At some point, he also worked as a barkeep. Not only immigrants, but poor and working-class New Yorkers rarely stayed in one job for very long. Rather, they bounced from one to the next, following the changing demands of the city's job market, which rose and fell according to the season. In the winter months, for example, when construction slowed, work was scarce for bricklayers, carpenters, and other laborers. For waiters, it was just the opposite. Winter was their boom season. But in the oppressive summer heat, when upper-class New York escaped to the shore or the mountains, city restaurants lost their clientele and waiters lost their jobs.

From the early nineteenth century onward, immigrants have played a vital part in feeding America. Working in jobs traditionally rejected by the native-born, they have peddled fruit, vegetables, fish, and thousands of other edible goods. They baked bread, slaughtered livestock, brewed beer, dipped candy, waited on tables, and cooked family suppers, to give just a few examples. Some of those immigrants—the Germans, for example—left a well-defined culinary footprint. On landing in New York, German immigrants established groceries, delicatessens, beer halls, lunch rooms, bakeries, and butcher shops, a kind of parallel food universe set apart from the city's existing network of food purveyors and tradespeople. All of this food-based activity satisfied the distinct culinary needs of the German community. At the same time, the shops and eating places, the sausage stands and sauerkraut vendors, were points of culinary transmission, places where a native-born American could sample his first grilled bratwurst, or pretzel or glass of lager. The Irish, however, never established that parallel universe. Instead, much like Bridget and

Joseph, they carved out a place for themselves in the existing culinary institutions of nineteenth-century America, a pattern that was rooted in their own culinary past.

When the potato, a New World food, first landed in Ireland at the end of the sixteenth century, the Irish larder was varied and nourishing, especially compared to the rest of Europe. At that time, the Irish diet was based on grain, mainly wheat, barley, and oats. Dried and ground into meal, these grains were baked into flat, dense griddle cakes, or boiled into a porridge called "stirabout." Dairy was another cornerstone food. The Irish drank buttermilk and skim milk, while butter was so plentiful it was eaten in chunks, the way we eat cheese. For protein, they had fowl, mutton, beef (both salted and fresh), bacon, ham, salmon, herring, eel, trout, and shellfish. Their gardens were planted with turnips, peas, beets, cabbages, and onions. The sea was another source of vegetable matter, supplying the Irish with mineral-rich water plants. Over the next few centuries, the Irish larder contracted sharply, and as it did, the potato moved from the sidelines to the center of the Irish diet, a shift linked to much broader changes in the political landscape.

Though English interference in Ireland dates back to the twelfth century, in 1649 Oliver Cromwell and his army changed the course of Irish history. In the summer of that year, Cromwell was dispatched to Ireland to suppress a Catholic rebellion, killing thousands of civilians in the process. To complete his program of retribution, Cromwell confiscated Catholic-held land, redistributing it to his Protestant supporters. From this point forward, English absentee landlords dominated the local economy, while the Irish were reduced to tenant farmers.

A bad situation now grew worse. A quickly growing population, combined with economic pressures of the landlord system, forced the Irish to live off smaller and smaller plots of land. With their farms shrinking, the Irish increasingly turned to the potato as their primary food source. As they knew from experience, the potato was relatively easy to grow, even in poor, rocky soil, but above all, it provided more life-sustaining calories per square acre than oats, barley, or any other crop. (They also knew that the potato plant was particularly sensitive to moisture and frost, but they conveniently brushed aside these less attractive qualities until it was too late.) So, by the late 1700s, the once-varied Irish diet was severely streamlined. The average farmer now consumed roughly seven pounds of potatoes a day, skim milk or buttermilk, and, occasionally, as Sunday treats, oat porridge or bacon and greens. The reliance on potatoes became even more pronounced over time. In the decade before the famine, potato consumption rose to twelve pounds a day for men and ten for women, supplying the farmer with most if not all of his calories.

The rural Irish lived on potatoes, but to earn rent money, they put aside a small patch of land for raising chickens, ducks, and turkeys, selling the eggs and, eventually, the birds as well. Some families kept a cow, churning butter from her milk, which they also sold, and keeping the by-product, buttermilk. But the farmers' most valuable asset was a more rotund and short-legged creature. Touring the Irish countryside, travelers to nineteenth-century Ireland reserved a page or two in their journals to marvel over the deference which rural families extended to their pigs. The following account comes from a particularly observant German visitor, Johann Kohl, who crisscrossed Ireland in the 1840s:

> *Nothing offers so striking a contrast to the meager, ragged wretchedness of the Irish peasant than the creature with which he shares his home—I mean his pig. . . . He feeds it quite as well as he does his children, assigns to it a corner in his sitting-room, shares his potatoes, his milk, and his bread, and confidently expects the pig will in due time gratefully repay.*[6]

That moment came when the coddled pig, "so oily, so round, so paunchy," was brought to market and sold, the profit going straight to the landlord's pocket. Half-jokingly referred to as "the gentleman who pays the rent," it was no wonder the pig was treated with such reverence. The family's survival depended on him.

On the larger estates, the Irish continued to grow their traditional cereal crops and to raise livestock, but not for home consumption. Rather, both were exported to foreign markets, including the United States, along with butter, ham, bacon, and other edible goods. So, while most of Ireland survived on a single food, warehouses in the major port cities were culinary treasure houses. Struck by the discrepancy, Herr Kohl, our German tourist, offers the following description of the wharf district in Cork:

> This city is well-known as the principal port for the exportation of raw produce of the whole of Ireland. . . . In the neighborhood of Cork are some of the greatest dairies in Ireland. Kerry and other cattle-grazing districts are also not very distant; so that here the largest quantities of butter, bacon, hams, meat, and cattle are brought together. . . . The quays of Cork present much that is interesting in all these varieties of merchandise, especially the embarkation of livestock, pigs, oxen, cows, etc. . . . One ship is being laden with firkins of butter for foreign lands, where Ireland must be thought one of the richest countries in the world, or she would not export those whole cargoes of fat.[7]

Contrary to popular conceptions of premodern Ireland as a food-starved nation, it was, in fact, exceptionally rich in edible resources. The peasantry, however, was excluded from the feast.

The Irish boiled their potatoes in large three-legged pots that stood over an open peat-fire. In between meals, women used their potato pots (in the west of Ireland, they were known as "bilers") for washing. After the meal, it was a trough for the family pig, who received the peels. Filled with brown bog water and placed outdoors, the biler doubled as a looking glass. To strain their potatoes, women used an oval-shaped basket called a "kish." Plates and cutlery were unheard of. Instead, women spread the potatoes on a table, if the family had one, and if not, on a clean patch of floor. In the absence of eating utensils, family members deftly peeled away the skin with their fingers to reach the steaming interior. Irish cooks satisfied their family's craving for variety with a category of food known as "kitchen," which was any fatty or highly seasoned morsel eaten along with potatoes, "to give the meal savor." The most elemental form of kitchen was an infusion of black pepper and water, consumed as a beverage. Some forms of kitchen—salt fish, bacon, pepper—required money. Other forms were free. Along the coast, Irish cooks gathered shellfish and various kinds of seaweed, which they dried, adding it to the potato pot as a seasoning.

The Irish referred to the famine of the 1840s as "the Great Hunger," to distinguish it from the many smaller "hungers" that had preceded it. The Irish historian William Wilde, father of Oscar Wilde, prepared a report for the 1851 Irish census on the many famines, crop failures, and related disasters in Irish history culminating in the Great Famine. Among them was the devastating cold snap of 1739–40 that caused nationwide destruction as the tubers froze solid into the ground. In fact, the verse that opens this chapter was written to commemorate the terrible loss of human life that followed.

No other immigrant arrived in the United States with a culinary tradition as skeletal as the Irish. By the time of the Great Famine, three centuries of the landlord system had stripped it down to a single carbohydrate and a handful of condiments. Where Germans and Italians and Jews worked hard to perpetuate native food ways, the Irish peasant had little to preserve. Other immigrant groups used their native foods to establish a collective identity in the New World. Not so the Irish. As Hasia Diner points out in her landmark book *Hungering for America*, the Irish found little to celebrate in the foods of their homeland. Rather, they turned to religion, music, drama, and dance, among other cultural forms, to assert their identity and connect themselves with the past and with each other. The one exception in all this, as Diner points out, was alcohol, mainly Irish whiskey. A symbol of Irish sociability and Irish independence, whiskey was the immigrant's only source of consumable pride. Still, the foods of home, no matter how meager, could haunt the immigrant—especially at life's most critical moments. The following encounter with a dying Irish seamstress was recorded by a New York charity worker in the 1860s. The woman, abandoned by her husband, lived with her two children in a tenement on the west side of Manhattan. "I was called in the other day and held a long conversation with her," the charity worker begins:

She has no more fears, or anxieties, she is not even troubled about
her little one. God will care for her. . . . She spoke of her many trials
and sorrows—they were all over, and she was glad soon to be at rest.

We asked about her food. She said she could not relish many things,
and she often thought if she could only get some of the good old plain
things she ate in Ireland at her brother's farm she should feel so much
better. I told her we would get some good genuine oatmeal cake from
an Irish friend. Her face lighted up at once, and she seemed cheered
by the promise.[8]

For our New York seamstress, the very plainness of Irish home-cooking
becomes its salient virtue. The same culinary aesthetic is at work in
the fiction of Seamus MacManus, an Irish-American writer popular
in the early twentieth century. The following excerpt comes from *Yourself and the Neighbours*, a sentimental look back at the simplicity of Irish
domestic life:

Oatmeal stirabout with lashings and leavings of buttermilk for break-
fast, and for dinner a pot of fine floury potatoes, that when spread
steaming on the table, were laughing thru their jackets to you to come
on. Sometimes Molly could afford you even a fine bowl of buttermilk to
kitchen the potatoes, and always plenty of salt, and pepper too.[9]

Even buttermilk! And plenty of salt! MacManus is clearly playing on the
meagerness of the Irish larder, showing his American readers how little
Molly had to work with, and yet how satisfying mealtime could be. The
"laughing" or "smiling" potato, an oft-used expression in nineteenth-
century Irish literature, hints at the deep affection the Irish reserved for
their most important food. For most, a steady regimen of potatoes and buttermilk was oppressively dull—a "diet fit for hogs" is how one Englishman
described it. But the Irish attachment to potatoes endured, even after the
Great Hunger, when the food they relied on had betrayed them so cruelly,
and continued among Irish immigrants in the United States.

Though food in the United States was more varied and abundant than it had been in Ireland, hunger was still a daily presence. After all, the bounty of the American market extended only to people with work, which, in bad times, could be elusive. Alice McDonald emigrated to the United States around 1866, during the rocky economic years directly following the Civil War. In a letter to her mother, Alice sums up her troubles. (The spelling and punctuation are her own.)

> *I take this opportunity of writing these few lines to you hoping to find you all in good helth as this leves me in at present . . . I wish I nevr came to New York it is a hell on earth . . . times is bad here for the last to years flour is 16 Dollars per barel beef 35 per pound . . . Butter by retell is seventy cents per pound you may think how the money has to go.*[10]

Their dream of a better life gone sour, struggling immigrants had nothing to look forward to, so they looked back instead, savoring the foods of a former existence. Mary Anne Sadlier, born in Ireland in 1820, was a popular nineteenth-century author of didactic or instructional novels for immigrant readers. This is how one of her characters, a near-destitute immigrant, compares America's scarcity to the good foods he remembers in Ireland:

> *Things are not so plentiful here . . . We haven't the big fat pots of bacon and cabbage, or broth that a spoon would stand in; no, or the fine baskets of laughing potatoes that would do a man's heart good to look at.*[11]

The hulking pots of cabbage and bacon may have never existed in precisely the way he remembers. In all probability, they were three-quarters empty, but in his memory the hungry immigrant was free to embellish his culinary past, filling the pots and thickening the broth. Cabbage and bacon and boiled potatoes were poor people's food, according to some, but to this homesick immigrant they represented the height of abundance.

The above excerpts notwithstanding, references to food in early Irish-

American writing are extremely hard to find. Like many immigrants, the Irish never imagined that their daily diet was worthy of documentation, and as it happens, the rest of America seemed to agree. Their potato-based cuisine was far less intriguing to outsiders than Italian pasta or German sausage, and journalists largely ignored it. As a result, the Irish immigrant kitchen is a world largely closed to us. However, a quantitative picture of the Irish-American cook can be found in the multiple cost-of-living studies published in the first decades of the twentieth century. Several were conducted by the United States Department of Labor to help establish fair working wages. The question of household budget and how it was dispensed was also taken up by academics of the era, including a sociologist named Louise B. More, who organized a detailed study of how working-class New Yorkers earned and spent their money. Over the course of two years, she followed two hundred families living in Greenwich Village, an ethnically diverse neighborhood with a hefty per-centage of Irish. Her report, *Wage-Earner's Budgets*, provides us with a rare glimpse of Irish-American home-cooking a generation after the Moores.

According to More, the Irish-American housewife controlled the family purse strings, spending more on food than any other item, including the rent. The study also tells us exactly how she stocked her kitchen. Here, for example, are the weekly food expenses for a family of ten:

Milk, 2 bottles a day $1.25
Eggs 50
Three cans of condensed milk for tea ("and to spread
on bread when the children have no butter") 27
One quart of potatoes a day at .10 a quart 70
Vegetables 70
Bread, 5 loaves at .05=.25 a day 1.75
One and one-half pound of
butter at .30 a pound 45
Jam, .05 a day, except on Sunday 30
One-half pound of tea, no coffee 20
One can of Baker's cocoa 18

7 lbs of sugar at .20 for 3½ lbs .... .40

Meat .25 a day I.75

Sundries .... .18

Total .... *$8.50*[12]

Potatoes, milk, oatmeal, and butter were core staples among the working-class Irish families, all foods familiar to the immigrant, though not perhaps in the same quantities. Cabbage, onions, and turnips, the few vegetables consumed by Irish-Americans, were familiar too. Other foods, however, were acquired in America, like bread and tea. But a greater change was the new abundance of meat. In Ireland, meat appeared on the dinner table several times a year, if that much. In America, mutton, beef, and veal became everyday fare, fried simply in a pan or cooked into stews for the evening meal. One woman in the study devised an ingenuous soup recipe based on salt pork, onion, and macaroni, an ingredient adopted from one of her Italian neighbors. An entire pot of it cost only 6 cents.

Another innovation: sugar. As the study shows, the Irish-American housewife went through several pounds a week, using it to sweeten the tea that she consumed more as a food than a beverage. Rich in caffeine, a cup of sweetened tea was an inexpensive way to stave off hunger between meals, while supplying the homemaker with the energy required to complete her daily tasks. Where Irishmen were notorious for their whiskey intake, Irishwomen were known for their tea habit—"tea inebriates" is how one doctor described them. The balance of the sugar went into the children's cocoa or was used for baking pies, cakes, and puddings, a skill the homemaker may have learned in her servant days. Below is a recipe for a "cheap pudding" from the *Irish Times*:

CHEAP PUDDING

Gather up all crusts and stale pieces of bread until you have sufficient for an ordinary-sized pudding; soak in water over-

night, then pour off and add two tablespoonfuls of flour or
rice, which goes to substitute for eggs; one ounce of raisins,
the same of currants; one tablespoonful of cinnamon or
whatever seasoning is preferred; add milk consistent, and
mix well. Put in a buttered bowl to steam for two hours; or
put it into a pudding dish to bake.[13]

Bridget and Joseph, along with the great majority of immigrants coming
to the United States in the second half of the nineteenth century, entered
the country through Castle Garden in Lower Manhattan. Originally built
as a fort, it was converted into an amusement hall in 1824 before its final
transformation into the nation's first immigrant landing depot in 1855.
The octagonal main hall and the complex of buildings surrounding it
were a full-service operation, complete with a foreign money exchange,
railway ticket office, hospital, quarantine, post office, showers, and restaurant. Here, immigrants got their first literal taste of America. It came
in the form of sausage, pie, bread (both brown and white), beer, and
cigars.

Each morning, the New York newspapers printed the names and
ETAs of ships scheduled to arrive at the Garden that day. Scanning the
daily announcements, immigrants living in and around the city could
meet their relatives in the Castle Garden reception room. Like an audience in a theater, the crowd in the reception room watched the unspooling
drama of one teary-eyed reunion after another. "It is certainly interesting
to witness these meetings," one observer notes:

> Here is the name of a comely Irish girl called out, she enters blushing
> and is the next moment in the arms of her faithful sweetheart, who
> left her home in Ireland three years ago, and has now sent for her to
> make her his bride. There is kissing and crying and squeezing, and
> applause from the bystanders, who for the moment forget that they
> themselves will probably do the same sort of thing.[14]

A scene very much like this (only the genders would have been reversed) may have taken place between Bridget and Joseph. Seventeen-year-old Bridget Meehan landed at New York's Castle Garden in 1863. Two years later, Joseph Moore, a twenty-year-old Dubliner, made the same trip. By the end of the year, the two were married and parents of their first of eight children. Judging from the accelerated sequence of events, Joseph and Bridget had met in Ireland, and were likely engaged at the time of Bridget's departure.

Like many newly landed Irish, they found lodging in Manhattan's crowded sixth ward. Their first daughter, Mary Catherine, was born at 65 Mott Street in the Five Points, a section of the Lower East Side famous for its high crime rate, decrepit housing, and unsanitary living

Jane Moore Hanrahan, circa 1900, daughter of Bridget and Joseph Moore.

Roger Joseph Hanrahan, a baggage clerk, and Jane Moore were married in New York in 1895.

conditions. (Outbreaks of cholera, typhus, and other deadly infections were commonplace, and mortality rates were higher in the Five Points than most other city neighborhoods.) Over the next ten years, Bridget gave birth to seven more children: Jane, Agnes, Cecilia, Theresa, Veronica, Josephine, and Elizabeth. Only four survived childhood.

After Mott Street, they lived at 150 Forsyth Street and remained there for a relatively extended stay, from 1866 to 1869. With three young daughters, they picked up again and moved into 97 Orchard Street in the heart of German *Kleindeutschland*.

Immigrants who arrived in the United States with no family to meet them spent their first days or weeks in a boardinghouse. To help new arrivals find a reputable establishment (many were not), an area had been set aside within Castle Garden, where licensed boardinghouse-keepers could solicit potential customers. Just beyond Castle Garden, immigrants met up with a much shadier class of boardinghouse representative. Known as "runners," they were scam artists of the first order, regularly vilified in the local press. To rope in customers, runners employed a handful of standard ploys: they would snatch the immigrants' baggage and offer to cart it—free of charge—anywhere in the city, delivering it to a local boardinghouse where it was immediately "put into storage." With his baggage held hostage, the immigrant was forced to spend the night, paying any fee the owner demanded. Or they would abscond with one of the immigrants' children, forcing the parents to follow. More insidious runners played on feelings of national kinship to establish a rapport with their marks and win their trust. Then they fleeced them.

In 1867, the Irish politician and newspaperman John Francis Maguire toured the United States to see how his countrymen were faring in their new home, and published his observations in a book called *The Irish in America*. To illustrate the depravity of boardinghouse runners, he included the following story, told to him by a great, broad-shouldered Irishman "over six feet in his stocking vamps." On landing in New York, the strapping Irishman was *[p]ounced upon by two runners, one seizing the box of tools, and the other confiscating the clothes. The future American citizen assured his obliging friends that he was quite capable of carrying his own luggage; but no, they should relieve him—this stranger, and guest of the republic—of that trouble . . . He remembered that the two gentlemen wore very pronounced green neckties, and spoke with a richness of accent that denoted special if not conscientious cultivation; and on his arrival at the boardinghouse, he was cheered with the announcement that its proprietor was from "the ould counthry, and loved every sod of it, God bless!"*[15]

A two-night stay at the boardinghouse cost the Irishman a small fortune, more than he would have paid for a sumptuous dinner at the Astor Hotel.

Immigrant boardinghouses were scattered through Lower Manhattan, but were concentrated near the wharves. They were especially thick along Hudson, Washington, and Greenwich Streets, because of their proximity to Castle Garden. Here, one could find boardinghouses with owners of every nationality, each patronized by their respective countrymen. No group, however, was better represented than the Irish. Not just in New York, but throughout urban America, Irish entrepreneurs opened boardinghouses and hotels, a business they had no particular background in but which answered a pressing demand for immigrant housing and which they learned "on the job."

In 1848, during the height of the Irish exodus, a self-promoting but charming Irishman named Jeremiah O'Donovan conducted a rambling tour of the eastern United States. The purpose of his trip was to hawk his masterpiece, a history of Ireland written in epic verse, a kind of Irish *Aeneid* with O'Donovan cast in the role of Virgil. O'Donovan kept a detailed travel log, and in 1864 it was published as a book. *A Brief Account of the Author's Interview with His Countrymen* is essentially a glorified record of every Irishman he meets on his American wanderings, with special attention given to those who bought his book. Like the well-bred host at a dinner party, he is gracious to a fault. O'Donovan cites every customer by name (there are thousands of them), describing each with a thumbnail dossier, complete with occupation and place of birth, followed by a long-winded account of the customer's most outstanding virtues. *A Brief Account* is exceedingly repetitive, but O'Donovan's intention is more than mere entertainment. Aside from confirming his own reputation, O'Donovan is engaged in a public relations campaign on behalf of Irish-Americans, his attempt to counter negative attitudes directed at the immigrant community.

Buried in O'Donovan's flood of adjectives are valuable kernels of information. For example, O'Donovan describes an extensive network of Irish "boardinghouses" and "hotels" (the distinction between the two is not always clear) that was already in place by the mid-nineteenth century. It covered the major East Coast cities like Boston, New York, Philadelphia, and Baltimore, extended west to Cincinnati and south to St. Louis, but also reached into more out-of-the-way places like Kittanning, Pennsylvania, and Lawrence, Massachusetts. O'Donovan also tells us something about the boarders. They included seamstresses, bakers, priests, medicine salesmen, boot-makers—all of them Irish. Finally, the author imparts a clear sense of the pressing need for room and board among a foreign population that was unusually young, most of the immigrants single and alone in a foreign country. Every boardinghouse along his winding itinerary is filled to capacity, the guests "as thick in numbers as swallows in a sand bank."[16] At many, O'Donovan is turned away for lack of space; others manage to shoehorn him in even if it means sharing a bed.

Many of the immigrant boardinghouses in New York had been carved from the city's most venerable old homes. Sturdy brick structures with sloping tiled roofs, they were built in the mid-eighteenth century by the merchant princes who dominated the colonial economy. Many had names familiar to us today: Rutgers, Monroe, Crosby, Roosevelt are just a few. Around 1750, one of those merchants, William Walton, started construction on a fabulous new estate along the East River. Built from the most expensive materials—English timber, German tiles, Italian marble, and rare tropical woods for its interior—the structure stood three stories tall with views of the water. The house remained in the Walton family for several generations, but over time the wide-open expanse that once surrounded it was filled in with warehouses, factories, and other commercial structures. The neighborhood lost its cachet, and Walton's descendants moved uptown to join the rest of fashionable New York. The once-grand interior was pillaged down to its shell. The ground floor was leased out to a German saloon-keeper, a plate-glass distributor, and a manufacturer of election flags. The main tenant, however, was Mrs. Connors, an Irish boardinghouse-keeper who rented the two upper floors.

A reporter who toured the old mansion in 1872 offers a quick glimpse into her kitchen, now housed in the former bedroom of the first Mrs. Walton. The original fireplace, he tells us, was ripped out and replaced with a monstrous black stove. Covered with various-size kettles and saucepans, it was presided over by the boardinghouse cook and her ever-bustling assistants. On the reporter's visit, the kitchen was redolent with the smell of boiling cabbage and burned ham, both typically Irish foods. For the most part, however, Irish boardinghouse cooking was governed more by issues of economy than national origin.

The American boardinghouse is today close to extinction, but for most of the nineteenth century and part of the twentieth it was a very common living option. Writing about New York in 1872, the journalist James McCabe portrayed the city as one vast boardinghouse, a description that could also apply to Philadelphia, Boston, and other large cities. City dwellers regarded the boardinghouse as a necessary evil, a kind of residential purgatory that one must endure before moving on to other forms of housing. The woman who stood at the helm of this much-disparaged institution, the boardinghouse landlady, became a stock character in the popular imagination of nineteenth-century urban America, best known for her cheapness. She made regular appearances in newspaper and magazine stories, often seen prowling the public markets for third-rate ingredients. Here she is shopping at one of the city's night markets, outdoor venues that catered to the budget-conscious:

> *That woman in the red reps underskirt, over-trimmed with velveteen,*
> *with very little bonnet but a jungle of artificial flowers and parterre*

of chignon, with almost the entire back of a gallipagos turtle in her
tortoise-shell earrings . . . with the coarse hand in the soiled gloves,
and the large greasy pocketbook tightly clasped—who is she? Her
index finger is uncovered. She uses it as a convenient prod to go
through meat and find out how tough it really is. Know her business?
Of course she does. A slatternly Irish girl, maid of all work, with a
basket big enough for a laboring man to carry, follows diligently at
her heels. "Beef," said the lady of the earrings very laconically, stop-
ping full before the butcher. He looks at her for a moment, and with
a long pole hooks down a piece of meat. It is a long, stringy portion
of singularly unappetizing appearance.[17]

Her purchase tells the story, in encapsulated form, of boardinghouse cuisine. As a rule, it was built on the most gristly cuts of beef,
mutton, and pork, organ meats, including heart, liver, and kidney,
salt mackerel, root vegetables, potatoes, cabbage, and dried beans—
the core ingredients of urban working people. Boardinghouse cooks
were known for their tough steaks, greasy, waterlogged vegetables,
and insipid stews. Their signature dishes—hash, soup, and pies—
were recycled from previous meals. The corned beef served for dinner
might reappear as the next day's breakfast steak, only to turn up again
in a meat pie or a hash.

Though hash in its many forms (not just corned beef) has drifted to
the culinary margins, at one time it was an American staple, especially
among working people. Along with meat pies, its raison d'être was to use
up leftovers. Economy-minded cookbooks of the period devoted whole
chapters to hash. Miss Beecher, for example, the well-known domestic
authority, gives eighteen hash recipes in her *Housekeeper and Healthkeeper*.
Among them are ham hash with bread crumbs, beefsteak hash with turnips, and veal hash with crackers. Though hash was found in home kitchens and cheap restaurants, it was most closely tied to the boardinghouse.
"Hash eaters," a common nineteenth-century label, was a synonym for
boardinghouse dwellers, while "boardinghouse hash" was a term used to
denote any mixture—edible or not—of uncertain composition. The rec-

ipe that follows is adapted from *First Principles of Household Management and Cookery* by Maria Parloa, who taught at the Boston Cooking School. The ingredients are inexpensive, but the end product is extremely satisfying—just what a good hash should be.

FISH HASH

> One half pint of finely chopped salt fish, six good-sized cold boiled potatoes chopped fine, one half cup of milk or water, salt and pepper to taste. Have two ounces of pork cut in thin slices and fried brown; take the pork out of the fry pan, and pour some of the gravy over the hash; mix all thoroughly, and then turn it into the fry pan, even it over with a knife, cover tight, and let it stand where it will brown slowly for half an hour; then fold over, turn out on the platter, and garnish with the salt pork.[18]

1836 was a watershed year in the culinary history of New York. Before that time, the city had relatively few public eating houses aside from taverns, which concentrated more on fluids than solids. Meals were generally consumed at home, with men returning from their place of work at midday, and again in the evening for supper. New York was much smaller then, so commuting twice a day was manageable, the distance between home and work no greater than a fifteen-minute walk. Public dining was reserved for the rich, or, paradoxically, the poor who took their meals at "coffee and cake shops"—a misleading name, since they also served pork and beans, hash, pies, and other low-budget dishes. Open twenty-four hours a day, their customers were laborers, newsboys, petty criminals, firemen, and other night workers, people of limited means with few eating options. Known for their "dyspepsia-laden cake," "oleaginous pork," and "treacherous beans," they were places to avoid if possible. As the city stretched northward, it began to divide itself amoeba-style into an uptown residential quarter and a downtown business district. Geographically cut off from their uptown homes, businessmen working in Lower Manhattan needed to be fed. In 1836, an Irishman named Daniel Sweeny opened a cheap eating house geared to the downtown worker of "medium class." Though prices were low (6 cents a plate), the food was decent and nourishing, the service efficient, and the surroundings clean. Sweeny's menu offered a fairly broad selection of nineteenth-century staples: oysters, roast beef, corned beef, boiled mutton, pork and beans, with pies and puddings for dessert. The restaurant was an immediate success, filling a gap in the city's culinary needs which no one before him had identified.

Sweeny's biography is emblematic of the striving immigrant. Born in Ireland in 1810, he emigrated to New York as a teenager, amassing a small fortune as a water vendor. (This was before the construction of the Croton aqueduct, when water was still hauled by bucket from a downtown reservoir.) He learned the restaurant business by working as a waiter. One of his customers was Horace Greeley, editor of the *New York Tribune*, who supposedly urged Sweeny to open his own eating house, which he did. With the money earned from his restaurant, Sweeny went on to larger, more ambitious projects, including the eponymous hotel that became a social hub for prominent Irish New Yorkers. Among his patrons were newspapermen, political bosses, and high-ranking members of the Catholic clergy. In the 1860s, when the hotel served as the unofficial New York headquarters for the Irish nationalist movement, Sweeny became host to revolutionary heroes like John Mitchell and Jeremiah O'Donovan Rossa. Sweeny staffed his establishments with immigrants like Joseph Moore, Irish waiters who banded together, often rooming in the same boardinghouse and helping one another find jobs.

Sweeny's experiment in public dining encouraged a number of imitators, many of them Irish. Patrick Dolan immigrated to New York in 1846 and learned the restaurant business working in Sweeny's hotel. When he turned twenty-five, Dolan opened a small lunch counter on Ann Street, which in time became a New York institution.

Dolan's Restaurant
            33 Park Row, New York City
Founded 1865

        Framed Bill of Fare on the Wall

            Copied 1. Jan. 1906
                            F. E. B.

    Coffee and Cakes   10
    Corned Beef   10
    Boiled Ham   10
    Pork and Beans   10
    Pickled Tongue   10
    Rice and Milk   10
    Oatmeal and Milk   10
    Bread and Milk   10
    2 Boiled Eggs   10
    Dry Toast   10
    Milk Toast   15
    Oyster Pie   10
    Oyster Stew   20
    Oysters, Raw   20
    Sandwiches   10
    Tea   5
    Coffee   5
    Pie   5
    Crullers   5

1906 menu from Dolan's Restaurant, an Irish-run lunchroom.

Dolan's Restaurant, which later moved to Park Row, was best known for two items. The first was "beef an," New York shorthand for a slab of corned beef and a side of beans. (A less popular option was "ham an.") The second was doughnuts, which New Yorkers called "sinkers." James Gordon Bennet, publisher of the *New York Herald* and a regular at Dolan's, was devoted to the oyster pie, as was Jay Gould, the railroad magnate. When Teddy Roosevelt returned to New York after his victory at San Juan Hill, he hired the Dolan's cook to prepare a dinner for four hundred of his Rough Riders. Alongside the illustrious, Dolan's fed the city's boot blacks and newspaper boys, people with limited budgets who demanded cheap, good food, served in a hurry. A typical lunchtime crowd at Dolan's was a startling mix of high and low, from street kids to robber barons, providing nineteenth-century New Yorkers with the kind of heterogeneous eating experience that no longer exists.

Joseph Moore most likely began his career in an Irish-owned restaurant similar to Dolan's. Raised in Dublin, he may have worked there as a waiter before emigrating to the United States, arriving in New York with valuable work experience. But even if he knew the job in a generic way, waiting tables in a New York restaurant required specialized knowledge. By the 1860s, New York waiters had evolved their own peculiar on-the-job dialect, shouting to the cook orders that mystified the average customer. Here are just a few of them:

> pair o' sleeve buttons – two fish balls
> white wings, end up – poached eggs
> one slaughter on the pan – porterhouse steak
> summertime – bread with milk
> Murphy with his coat on – boiled potato, unpeeled
> solid shot – apple dumpling
> shipwreck – scrambled eggs
> mystery – hash
> one – oyster pie[19]

Along with oyster pie, the cheap eating house offered oysters on the half shell, fried oysters, broiled oysters, oyster stew, oyster pan roast, oysters Baltimore-style, and oyster omelet. A dish called oyster patty, a prebaked puff pastry shell filled with oysters in gravy, was created on the spot for waiting customers by the "patty man," a restaurant worker with a specialized skill. Watching him at work, if he was especially talented, was a form of free entertainment. One New York patty man, an exceptionally nimble Irishman, caught the attention of a local reporter, who

A mid-nineteenth-century New Yorker enjoying freshly shucked oysters from an Irish-owned oyster stand.

deconstructed his manipulations into eight quick, but distinct motions, all performed with consummate grace.

The recipe below is from *Jennie June's American Cookery Book*, published in 1870:

OYSTER PATTIES

Beard the oysters, and, if large, halve them; put them into a saucepan with a piece of butter rolled in flour, some finely shred lemon rind, and a little white pepper, and milk, and a portion of the liquor from the fish; stir all well together, let

it simmer for a few minutes, and put it in your patty pans
[resembling shallow cupcake pans], which should already be
prepared with puff paste in the usual way. Serve hot or cold.[20]

New York was growing in size, volume, and stature, assuming its place
as a world-class commercial capital. Eating places like Sweeny's answered
the changing culinary needs of a city in transition. At the same time, it
literally fed the engine of change. The great nineteenth-century reporter
George Foster sums up the dynamic with his customary flair:

> We are inclined to believe that, notwithstanding their steaming-rooms
> and thin soups, it is to the Eating Houses that New York is in a great
> measure indebted for that continuous rush of commercial activity
> around her great business centres, which so strikingly distinguishes
> her from all other cities . . . Just think of it—two or three thousand
> people going up and down the same stairs and dining at the same
> tables, within three hours! Such a scene cannot be imagined by any
> but a New Yorker. Nowhere else, either in Europe or America, does
> anything like it exist. It is the culmination, the consummation, the
> concentration of Americanism.[21]

Apparently, Foster was unaware that this quintessentially American institution had its roots in the driving creativity of immigrants.

When Bridget and Joseph landed in New York in the 1860s, the food
most closely identified with Irish-Americans was already a gastronomic
fixture on this side of the Atlantic. Though corned beef and cabbage
traveled to the United States with the first Dutch settlers in the early
seventeenth century, many of the groups that followed, including the
English, the Germans, and the Jews, emigrated with their own corned
beef traditions.

Corned beef belongs to a large family of preserved meats and fish.
Many, like smoked salmon, are now considered delicacies, but a cen-

tury or more ago they were foods of necessity. The invention of modern refrigeration, starting with the icebox in the early nineteenth century, alleviated one of the cook's most vexing challenges: keeping food relatively fresh and edible. To meet this critical need, women across Europe devised several techniques, each of them based on one, or sometimes two different preserving agents. Those agents were heat, smoke, salt, and acid. So, for example, meats, fish, and fowl were generally smoked, salted, or pickled, while fruits and vegetables were pickled, jarred, or dried.

The term "corned beef" refers to the large grains, or corns, of salt with which the meat was cured. An early account of how this was done appears in a British domestic manual from 1750, *The Country Housewife's Family Companion*:

After the beef has been sprinkled with salt, and lain to drain out its bloody juice six or seven hours, wipe every piece dry, and rub them all over well with dry hot salt. This done, pack them close in a pot or tub one upon another.[22]

Housewives performed the same basic procedure every year around harvest time, or whenever cattle were slaughtered. In the cities, meanwhile, urban dwellers could buy their beef already salted from local food purveyors. A generous portion of it was produced in Cork, Ireland. Along with butter, bacon, cheese, pickled meats, and preserved fish, Irish food manufacturers exported thousands of barrels of corned beef to countries throughout Europe. Some traveled even farther, crossing the Atlantic to European settlements in the New World. Cork provisioners shipped immense quantities of corned beef, along with other edible goods, to the West Indies, where farmland was reserved for the lucrative business of growing sugar cane. Up until 1776, some of those goods, corned beef included, were exported to the English colonies in North America.

Wherever corned beef was eaten, cooks developed ways of preparing it that harmonized with local food traditions. In Germany, it was thinly sliced and served with black bread. Scottish cooks combined their corned beef with a regional staple, oats. First, they simmered the meat along with carrots, parsnips, potatoes, cress, and cabbage. When the vegetables were tender, they added a scoop of oatmeal to the pot to thicken the stock. English cooks included corned beef in their savory pies. Just across the North Sea, the Irish improvised their own corned beef creations; one was boiled corned beef and cabbage. In 1854, an information-packed article on Dublin street vendors appeared in an English journal called *Ainsworth's Magazine*. The author, an Irishman named Matthew Lynch, lists the many food peddlers who plied their wares. Oysters, cockles, herrings, turnips, peas, cauliflower, strawberries are some of the foods he mentions. Another common item was cabbage, which, according to Lynch, was scarcely ever eaten unless accompanied by either bacon or corned beef. "A rump of beef and cabbage is a favorite dish with all persons in Ireland—either peers or peasants," Lynch informs us. Maybe this was

Businessmen mingle with street urchins in a downtown lunchroom. Freshly carved corned beef (far left) was a lunchroom staple.

true in an idealized Ireland, but in reality corned beef was beyond the means of the average farmer. In fact, many of the poor Irish who arrived in the United States after the 1840s had likely never even tasted it. Much more likely, the traditional Irish pairing of corned beef and cabbage had been carried to the United States by well-off merchants and industrialists, most from the Protestant north of Ireland, who had settled here a century earlier.

In nineteenth-century America, corned beef and cabbage developed a split personality. At the cheapest New York lunchrooms, a serving large enough to feed a family of five was ladled into tin buckets (an early version of "takeout") and sold for 15 cents. Corned beef also gave sustenance to the city's newsboys, a population of orphaned kids who lived on the streets, supporting themselves by hawking the daily papers. Beginning in the 1850s, newsboys were able to buy a bed for the night at one of the city's Newsboys' Lodging Houses, a New York charity that remained active well into the twentieth century. In addition to shelter, the boys received two meals a day. For breakfast, they had coffee, oatmeal, and bread and butter, and for supper, a rotating selection of cost-effective entrées. Here is one typical dinner lineup from 1895: "Sunday, roast beef; Monday, pork and beans; Tuesday, beef stew; Wednesday, corned beef and cabbage; Thursday, pork and beans; Friday, fish balls; and Saturday, pork and beans."[23] Inexpensive and easy to cook, corned beef and cabbage was a staple of the institutional kitchen. Along with orphanages and military camps, it also made steady appearances in hospitals and prison mess halls.

But corned beef and cabbage also led a much more exalted existence, featured in some of the nation's most exclusive dining venues. Contrary to our romantic projections of the nineteenth-century family Christmas, wealthy New Yorkers frequently spent the holiday in a hotel dining room. In anticipation of Christmas Eve, hotel chefs across the city composed the most lavish multicourse dinners imaginable, competing with one another in a kind of unofficial holiday cook-off. On Christmas Day, their menus were published in the local papers. The *New York Times* Christmas dinner roundup for 1880 began:

> *The discriminating palate of the sybarite was necessary for a full ap-*
> *preciation of yesterday's Christmas dinners at most of the big hotels*
> *in this cosmopolitan town. At the Windsor Hotel, one of several*
> *mentioned, the menu was all that a gourmet could ask for.*[24]

Among the offerings were Maryland terrapin, canvasback duck with cur-
rant jelly, *and* corned beef and cabbage. At a New Year's Day luncheon
hosted by the Fifth Avenue Hotel, corned beef and cabbage shared the
table with pâté de foie gras, tournedos of beef with sauce béarnaise, and
sweetbreads *financière*.

While the well-to-do native New Yorker celebrated with corned beef
and cabbage, the Irish had disowned their ancestral food. Beginning in
the late eighteenth century, each year on St. Patrick's Day, Irish-American
societies convened for all-male holiday banquets, formal and highly struc-
tured events that began with a seven-course meal and ended with a series
of very long toasts. The banquet room was customarily decorated with
Irish and American flags, portraits of St. Patrick (some executed in sugar
paste, others drawn in wax on the mirrors), emblems of harps and sham-
rocks. The bulk of the menu, however, comprised the same Frenchified
foods served at any New York banquet, with perhaps a single symboli-
cally Irish food thrown into the mix as a kind of accent. So, for instance,
along with *sole farcie au vin blanc*, and filet of beef, you might find "potatoes
served in their jackets," or Irish bacon with greens. Like other immigrants,
New York's Irish elite faced the tricky task of straddling two cultures, one
rooted in their Irish past, the other in their present lives as assimilated
citizens. (The same identity-juggling is reflected in the banquet menus
of the well-to-do German *vereine*. Gathered in their meeting halls, the
banquet room decorated with German and American flags, *vereine* members feasted on Kennebec salmon with sauce Hollandaise, chicken *à la
Reine*, and potatoes *Parisienne*, foods that represented the culinary cutting
edge. In acknowledgment of their ethnic roots, however, the menu also
included a handful of token German specialties, perhaps asparagus with
Westphalian ham, or chicken soup with marrow-filled dumplings.)

Through a gradual, haphazard process, second- and third-generation immigrants reclaimed corned beef and cabbage as a quintessentially Irish food. One early instance of this culinary appropriation can be found in a comic strip. "Bringing Up Father," created by George MacManus in 1913, stars Maggie and Jiggs, an Irish immigrant couple suddenly thrust into high society. How each of them responds to their newfound wealth provides the comic's basic story line. Maggie is a dedicated social climber, determined to wash away all traces of her working-class roots, while Jiggs is happy to go on carousing with his pals as if nothing has changed. A continuing point of tension between them is Jiggs's unshakable affection for corned beef and cabbage, the food of his hardscrabble past. Below is one example of their many corned beef–inspired spats:

Maggie and Jiggs's high jinks inadvertently launched a chain of gastronomic events that transformed both the perception and reality of Irish-American food ways. In 1914, a New York saloon-keeper, James Moore, opened a restaurant on West 47th Street, in the city's emerging theater district. In homage to Maggie and Jiggs, he called it Dinty Moore's, the comic-strip tavern to which Jiggs is always sneaking off. Like the fictional version, the real-life Dinty Moore's served homey cooking, but the crowd it attracted was distinctly upscale; theater and publishing luminaries, politicians, and, in later years, broadcasting executives. Having grown up in a rough-and-tumble Irish neighborhood, James Moore was catapulted into New York society, but, like Jiggs, he held tight to his culinary roots, proudly known as "the corned beef and cabbage king," a dish that remained on his menu until his death in 1952.

Thus, in two very public forums—the comic strip and the restaurant—corned beef and cabbage became reattached to its Irish past.

When Joseph and Bridget Moore lived at 97 Orchard Street, they would have marked St. Patrick's Day with, perhaps, a dish of pig jowls, a common celebratory food among Irish East Siders. By the 1940s, during the lifetime of their great-grandchildren, corned beef and cabbage had become the mandatory St. Patrick's Day meal. At the same time, the era's food authorities denounced it as a culinary myth—a food pretending to be Irish that wasn't Irish at all. To explain its prevalence among Irish-Americans, they invented a number of historic scenarios. In the one most oft-repeated, Irish-Americans were introduced to corned beef, a food unknown to them in Ireland, by their Jewish neighbors and adopted it as a cheaper substitute for their beloved bacon.

The story of corned beef illustrates a larger point: immigrants used food as a medium to express who they were and who they wanted to become. They used it to assert identity, and in some cases to deny it. In the nineteenth century, socially prominent Irish-Americans, like our well-heeled banquet guests, found it expedient to distance themselves from certain low-status foods. But as the immigrants' social position evolved and anti-Irish feelings began to fade, the significance of their foods evolved too. The language of food, like any expressive medium, is never fixed but perpetually a work in progress.

CHAPTER THREE

The Gumpertz Family

O ur visit with Mrs. Gumpertz begins on a Friday, late morning, over a steaming pot of fish, a carp. The fish lays snugly in an oblong vessel, like a newborn in a watery cradle. From our current vantage point, it looks intact. In reality, however, the fish has been surgically disassembled and reassembled. It is the kind of culinary operation worthy of the trained professional, yet the responsible party is standing in front of us, an ordinary home cook. The process begins with a slit down the backbone. Mrs. Gumpertz opens the fish the same way one opens a book. Carefully, she scrapes the flesh from the skin, chopping it fine so it forms a paste, what the French call a forcemeat. Reduced to a mere envelope, head at one end, tail at the other, it is now the perfect receptacle for stuffing. Mrs. Gumpertz fills the skin with the paste and sews it shut. She lays the reconstructed carp on a bed of fish bones and onion—sliced but unpeeled—then puts it up to simmer. Just now, she is standing over the open pot, wondering if it needs more time. She prods it with a spoon; the fish is ready. She lifts the pot from the stove, moves it to a chair in the parlor, and leaves it there to cool by an open window. Moments before sun-

IMMIGRANTS LANDING AT CASTLE GARDEN.—DRAWN BY A. B. SHULTS.

Jewish immigrants landing at Castle Garden, 1880.

down, start of the Jewish Sabbath, she slices her carp crosswise into ovals and lays them on a plate. The cooking broth, rich in gelatin from the fish bones, has turned to jelly. The onion skin has tinted it gold. Mrs. Gumpertz spoons that up too, dabbing it over the fish in glistening puddles. To a hungry Jew at the end of the workweek, could any sight be more beautiful?

Twenty-two-year-old Natalie Reinsberg emigrated to New York from Ortelsburg, Prussia, in 1858. Her betrothed, Julius Gumpertz, another German Jew, had arrived a year earlier. Their wedding date is unknown, but their first child, Rosa Gumpertz, was born in New York in 1867. A second daughter, Natalea, known to the family as Nannie, was born in 1869, then Olga in 1871. The couple also had a son named Isaac, born in 1873, but he did not survive childhood. The family moved to 97 Orchard Street in 1870, when most of the building's residents were German-born Catholics or Protestants, and remained on

Orchard Street for the next fifteen years, as the neighborhood around them gradually shifted from Gentile to Jewish.

For Jews like the Gumpertzes, the Friday evening meal was reserved for fish, a tradition carried over from Europe. On the Lower East Side, the Sabbath fish tradition brought a stream of basket-wielding shoppers to the intersection of Hester and Norfolk streets, center of the Jewish fish trade in the 1890s. By this time, Hester Street was a full-blown pushcart market open every day except Saturday. The real action, however, began Thursday afternoon and peaked Friday morning, when Jewish women did their Sabbath marketing. This was prime time for the East Side pushcart vendor. Nineteenth-century New Yorkers who ventured downtown from the better neighborhoods above Fourteenth Street were flabbergasted by the scene awaiting them on market day: "There is hardly a foot of Hester Street that is not covered with people during the day. The whole place seems to be in a state of perpetual motion and the occasional visitor is apt to have a feeling of giddiness."[1] At the corner of Norfolk Street, the shoppers reached maximum density, a solid throng

An illustration from Harper's Weekly *depicting the Hester Street pushcart market, 1884.*

of housewives sorting through wagons of perch, whitefish, and carp for the freshest, clearest-eyed specimens. But now we're jumping ahead, beyond the scope of our present story. . . . Back in the 1870s, when the Gumpertz family moved to Orchard Street, East Side women bought their provisions from the public market on Essex Street or, perhaps, from one of the roving peddlers—some with baskets, others with wagons or carts—who patrolled the streets of Manhattan.

The Friday evening fish recipe was determined by where exactly the immigrant was born. If she came from Bavaria, for example, the housewife stewed the fish in vinegar, sugar, a splash of dark beer, and a handful of raisins, the sauce thickened by a sprinkling of crumbled ginger snaps. This was the famous sweet-and-sour dish known on Gentile menus as carp, Jewish-style. Another possibility was carp in aspic. Here, the whole fish was cut into steaks, simmered with onion and bay leaf, then allowed to cool with its cooking broth. The choicest portion was the head, appropriately reserved for the head of the household. Or perhaps, if she had an expanded food budget, the German cook might prepare an aromatic stewed fish, the sauce enriched with egg yolk. Such recipes were memorialized in *The Fair Cook Book*, a collection of German-Jewish recipes published in 1888 by the women of Congregation Emanuel in Denver, Colorado. *The Fair Cook Book* is the first known Jewish charity cookbook published in America (the queen of the genre, *The Settlement Cook Book*, has sold over two million copies to date). The following recipe, contributed by Mrs. L. E. Shoenberg, a nineteenth-century Denver homemaker, combines the sweetness of ginger and mace, the creaminess of egg yolk and the piquancy of lemon:

Stewed Fish

Cut a three-pound fish in thick slices and put on to boil, with one large onion sliced; salt, ginger, and mace to taste; cold water enough to cover fish, let boil about twenty minutes; take the yolks of three eggs, beat light, juice of

two lemons, chopped parsley, beat well together. When fish
is done pour off nearly all the water, return to fire and pour
over your eggs and lemon, moving fish briskly back and
forth for five minutes so that the egg does not coagulate.[2]

But if the cook was a native of Posen in eastern Prussia, the Friday night
fish might resemble Mrs. Gumpertz's carp. It is the dish we know today,
though in an altered form, as gefilte fish. The name "gefilte fish" comes
from the German word *gefülte*, meaning stuffed or filled, since the original
version was exactly that, a whole stuffed fish. Writing on the provenance
of gefilte fish in the 1940s, the Jewish cooking authority Leah Leonard
posed several possibilities:

> *Gefilte Fish may have originated in Germany or Holland sometime
> after the expulsion of Jews from Spain in 1492. Or it may have
> been invented in Russia or Poland. Or, perhaps, it was only the
> culinary ingenuity of a housefrau-on-a-budget in need of a food
> stretcher. One thing is certain, Gefilte Fish is Jewish.*[3]

Across Central and Eastern Europe, one could find some version of
gefilte fish wherever Jews had settled, prepared, like clockwork, Friday
mornings, and served that evening with grated horseradish. Aside from
matzoh or challah, few Jewish foods were as ubiquitous. Here was a food
of towering stature in the Jewish imagination. Over the centuries, a body
of mystical thinking had grown up around gefilte fish, explaining its perfection as a Sabbath delicacy. Because of its intricacy, the dish was also a
perfect measure of the Jewish housewife's culinary skill. No other food
in the Jewish kitchen required as much time or finesse. Along with the
Sabbath candlesticks, the oblong gefilte fish pot, a vessel dedicated to
that one food, was among a handful of objects that the Jewish housewife
carried with her to America.

Despite its Jewish resume, gefilte fish did not originate with the Jews.

Rather, it was a culinary convert, a food taken from the Gentile kitchen and adapted by the Jewish cook sometime in the distant past. In this respect, it reflects a larger pattern true of many foods typically consumed by Jews, among the world's most avid culinary borrowers. Where most cuisines are anchored to a place, Jewish cooking transcends geography. Spatially unmoored, it is the product of a landless people continuously acquiring new foods and adapting them as they move from place to place, settling for a time, then moving again.

Coming from Prussia, Mrs. Gumpertz was an Ashkenazi, a very elastic label that takes in the Jews of northern France, Germany, Austria, Romania, Poland, all of the Baltic countries, and Russia. Its original meaning, however, was more narrowly defined. Sometime in the tenth century, large Jewish families from southern France and Italy began to migrate north, forming settlements along the Rhine River. These were the original Ashkenazim, a term derived from the medieval Hebrew word for Germany. The early Rhineland communities were made up largely of rabbis and merchants. Both figures, it turns out, played major roles in shaping Ashkenazi food traditions. In the great centers of Jewish learning that sprang up in the Rhine Valley, rabbinic scholars directed their intellectual energies toward food-based issues, including the finer points of kashruth, Jewish dietary law. As interpreters of kashruth (which is ever-evolving), they decided which foods were fit for Jewish consumption, how they should be cooked, who was allowed to cook them, and when they should be eaten. Jewish traders, meanwhile, acted as culinary conduits, shuttling foods and food traditions from one side of the globe to the other. As the preeminent travelers of their day, they introduced medieval Europe to the exotic foods of the East: nuts, spices, marzipan, and, most important of all, sugar. On a smaller geographic scale, they carried foods from town to town and country to country, spreading localized food traditions within Europe and creating regional cuisines.

The flow of Jews from southern Europe (most were from Italy, where Jews had been living since the days of the Roman Empire) continued through the twelfth century. By this time, a distinct Jewish culture had evolved in the Rhineland and taken root, but only temporarily. The ever-shifting political environment kept the Jews moving. The period of the Crusades, which began at the end of the eleventh century and lasted for another two hundred years, was a particularly difficult period for the Ashkenazim. On their way to the holy land, crusading soldiers, in a fit of religious zeal, would stop to torture Jews, in some cases wiping out entire towns.

Jewish hatred stirred up by the Crusades set the tone for the next several centuries. State-sponsored expulsions, massacres, and anti-Jewish riots pushed the Jews farther east and north into Poland, Lithuania, and beyond. At the same time, more subtle forms of persecution prevented Jews from staying too long in any one place. Within German-speaking Europe, locally enforced laws restricting the Jews' right to own property, to work in certain occupations, to live where they chose, and even when they could marry left the Jews both rootless and poor. Many worked as itinerant peddlers, traveling by foot and selling assorted dry goods, pots and pans, needles, thread, and fabric. The truly destitute lived as wandering beggars. For the most part, the Jewish migrations flowed eastward, but if the political situation in Poland or Russia became too inhospitable, Jews circled back into Germany.

The history of Ashkenazi cooking tells the story of a people in motion. Since they came from Italy, it shouldn't surprise us that many early dishes show a strong Italian influence. The most obvious is pasta, or noodles, which the Jews called *vermslich*, or *grimslich*, words derived from the Italian "vermicelli." In one medieval noodle dish, a favorite among twelfth-century rabbis, the dough was cut into strips, baked, and drizzled with honey, an early ancestor of noodle kugel. Boiled noodles arrived in Germany roughly three centuries later, another food carried north, this time by traders, many of whom were Jews. In his book *Eat and Be Satisfied*, John Cooper describes a dish called *pastide*, an enormous meat pie of Italian origin, typically filled with organ meats. Too large to finish in a single sitting, the pie was baked in its own edible storage container: a thick whole-grain crust that was chipped away at each successive meal. Like noodles, *pastide* was generally eaten on Friday evenings, a Sabbath tradition that lasted through the eighteenth century.[4]

While the Ashkenazi cook retained elements of her Italian past, she also adopted local food habits, creating a new hybrid cuisine. Like her Gentile neighbors, she relied on dried peas and beans, porridge made from millet and rye, black bread, cabbage, turnips, dried and pickled fish, and, eventually, potatoes, a nineteenth-century addition to the Jewish diet. A shared dependence on local resources created broad similarities between the two kitchens, Jewish and Gentile. More interesting, however, is the cross-over of specific dishes, a process helped along by the Jewish merchant, a crucial point of contact between the two cultures. Among the dishes that made that journey are two Jewish staples. Before the Jews adopted it, the braided bread we know as challah was the special Sunday loaf of German Gentiles. German Jews adopted the braided bread,

Even the poorest Jews celebrated the Sabbath with challah, the traditional braided loaf. Here, an immigrant prepares for the Sabbath in a Ludlow Street coal cellar, 1900.

which was originally made from sour dough, and renamed it *berches*, a term derived from the Hebrew word for "blessing." On the Sabbath table, the *berches* symbolized the offerings of bread once made to the *Kohanim*, the priests who served in the ancient temple. When a piece of bread was torn from the loaf and dipped in salt, it referred back to the sacrifices of salted meat at the temple altar.

Challah offers just one example of how borrowed foods could be reborn, their former lives erased from memory. Gefilte fish followed a similar path. The idea of a reassembled fish comes straight from the imagination of the medieval court cook, a master of visual trickery. Descriptions of medieval banquets are brimming with all manner of reassembled animals, from deer to peacock, brought to the table in their original skins. Following in that same tradition, gefilte fish was a creation of the court cook intended for the aristocratic diner. Here, for example, is a recipe from a sixteenth-century cookbook by the German court cook, Marx Rumpolt:

Stuffed Pike

Scale the fish and remove the skin from head to tail; cut the meat off from the bone and chop it fine with a bit of onion; add a bit of pepper, ginger, and saffron, also fresh, unmelted butter and black raisins, egg yolk, and a bit of salt. Fill the pike with this mixture, replace the skin, sprinkle on some salt, place it in a pan and roast it. Make a sweet or sour broth under it and serve either warm or cold.[5]

Recipes very similar to Rumpolt's can be found in German-Jewish cookbooks from the mid-nineteenth century. As it traveled from one culture to another, gefilte fish, much like challah, was invested with a new iconography. But where challah looked to the Biblical past, gefilte fish became a symbol of the messianic banquet awaiting the Jews in paradise where, according to the Torah, the righteous shall dine on the flesh of the Leviathan. On the Sabbath table, gefilte fish was the Leviathan, that giant sea creature, a taste of paradise on earth.

In the second half of the nineteenth century, East Side Jews like Mrs. Gumpertz continued the gefilte fish tradition, preparing it in the old style, much like Rumpolt had four centuries earlier. A few decades later, a folksier version of gefilte fish seems to have taken its place. Prepared by cooks from Poland, Lithuania, and Russia, the chopped fish mixture was simply rolled into balls, simmered, then served cold with horseradish, the all-purpose Jewish condiment. With Jews from disparate countries all gathered in one neighborhood, subtle regional variations suddenly took on significance. Polish Jews, for example, seasoned their gefilte fish with sugar, where Lithuanians favored pepper. East Side Jews saw the sugar/pepper divide as a token of larger cultural differences between the Galiciana (Polish Jews) and the Litvaks (Lithuanians and Latvians), using it in conversation as a kind of code. So, if an East Sider wanted to know what part of the world a fellow Jew came from, he could ask, half-jokingly, "How do you like your gefilte fish, with sugar or without?"

Here's a classic version of gefilte fish from the *International Jewish Cookbook*, a dish of surprising delicacy to anyone who has tasted the mass-produced version found in the kosher aisle of your local supermarket:

GEFILLTE FISCH

Prepare trout, pickerel, or pike in the following manner: After the fish has been scaled and thoroughly cleaned, remove all the meat that adheres to the skin, being careful not to injure the skin; take out all the meat from head to tail, cut open along the backbone, removing it also; but do not disfigure the head and tail; chop the meat in a chopping bowl, then heat about a quarter of a pound of butter in a spider, add two tablespoons chopped parsley, and some soaked white bread; remove from the fire and add an onion grated,

salt, pepper, pounded almonds, the yolks of two eggs, also
a very little nutmeg grated. Mix all thoroughly and fill the
skin until it looks natural. Boil in salt water, containing a
piece of butter, celery root, parsley, and an onion; when
done, remove from the fire and lay on a platter. The fish
should be cooked for one and one-quarter hours, or until
done. Thicken the sauce with yolks of two eggs, adding a
few slices of lemons. This fish may be baked but must be
rolled in flour and dotted with bits of butter.[6]

By the mid-nineteenth century, a distinct form of Jewish life had evolved
in East Prussia, the region where Natalie Gumpertz spent her first
twenty-odd years. The Jews here were thinly scattered in small towns and
villages, representing only a tiny fraction of the local population. Out of
two million East Prussians, fourteen thousand were Jews. Dispersed as
they were, East Prussian Jews lacked the critical mass to sustain the kinds
of Jewish institutions found farther east in the larger, more-bustling Polish shtetlach. The town of Ortelsburg, East Prussia, where Natalie was
born, was a sleepy market town, its Jewish population never much larger
than that of a single East Side tenement. Too small and too poor to
support a Jewish school, in the 1840s, the Ortelsburg Jews pooled their
resources to build a synagogue, but never hired a permanent rabbi. No
rabbi, no school. And yet, this remote outpost of European Judaism was
of sufficient size to accommodate two Jewish-run taverns.

Among rural Jews, the local tavern was *the* preeminent social spot,
especially for men, who came to play cards (a popular Jewish pastime),
read the newspaper, and drink. Over a mug of beer or glass of schnapps,
Jewish businessmen, from shopkeepers to horse dealers, cemented part-
nerships, found new customers, and made new contacts, not only with
fellow Jews but with Christians too, who were likewise tavern customers.
In many cases the taverns were attached to roadside inns that catered to
Jewish merchants and traders. The inn provided them with a bed to sleep
in and a stable for their horses, while the tavern kitchen, usually run by

the tavern keeper's wife, provided them with sustenance. On Saturday afternoons, whole families would stop by the tavern for a late lunch, paying for it when the Sabbath was over. At the start of the nineteenth century, the vast majority of German Jews lived in the countryside, but in the cities, too, Jews found their way into the hotel and restaurant business, one facet of their larger role in the food economy. Set apart from the wider culture by their distinct food requirements, the Ashkenazim relied on a vast network of butchers, bakers, vintners, distillers, traders, and merchants. The tavern-keeper belonged to this culinary workforce, supplying Jews with kosher food and drink in a public setting. Remember, too, that Christian-owned establishments were free to turn Jews away and often did. Jewish taverns and cafés, hotels, restaurants, and even Jewish spas, were the answer to widespread discrimination.

In the mid-nineteenth century, German Jews brought their experience in the hospitality business to America. At Lustig's Restaurant in New York, nineteenth-century Jewish businessmen dined on German specialties like sweet-and-sour tongue, stuffed goose neck, and almond cake, all prepared to the highest kosher standards. Nicknamed the Jewish Delmonico's, the Mercer Street restaurant stood in the heart of the old dry-goods district. Restaurant patrons were observant Jews from around the city, mainly wealthy merchants, who congregated at Lustig's for the refined kosher cooking, but also for the traditional ambience.

In the Lustig's dining room, with its bare wood floor and simple furnishings, customers performed the food-based rituals inseparable from the traditional Jewish dining experience. Like a visiting anthropologist, a New York reporter wrote up his observations of the lunchtime scene:

> As each customer came in, he would take off his overcoat, hang it up, and then go to the washstand and wash his hands, looking very devout in the meantime and moving his lips in rapid muttering. He was repeating in Hebrew this prayer: "Blessed be Thou, O Lord our God, king of the universe, who hast sanctified us with Thy command, and has ordered us to wash hands." Having thus performed his first duty, he took his seat and ordered his dinner.

The close of the meal was marked by the same kind of singsong praying that had opened it. In between, diners performed other curious acts, like dipping their bread into salt and praying over that as well. Also odd was how none of the customers removed their hats, a clear breach of dining etiquette. All in all, the experience was so foreign that the reporter exited Lustig's feeling disoriented, like "a traveler returned from strange lands to his native heath."[7]

When he left his office and boarded the uptown streetcar, the typical Lustig's customer returned to a traditional Jewish life. Beginning in the fall with Rosh Hashanah, the Jewish New Year, he celebrated the full calendar of holidays, and every week observed the Sabbath. Friday evenings he went to synagogue, and Saturdays too, returning home at midday for lunch and a nap. The rest of the afternoon he devoted to Torah. While the man of the house prayed and studied, responsibility for guarding the purity of the home fell to his wife. A task that revolved around food, the work was relentless. Training began in childhood, when she was old enough to stand on a chair in her mother's kitchen and help pluck the chicken or grate the potatoes. Over the years, she absorbed an immense store of food knowledge, allowing her to one day take control of her own kitchen. This happened the day she was married.

The traditional Jewish homemaker inherited her mother's recipes along with her expert command of the Jewish food laws. Each day of her married life, as she shopped for groceries, or cooked, or even cleaned, she called on her knowledge of kashruth, ensuring every morsel of food served to the family was ritually pure. In other words, kosher. Such cooks could be found among New York's German Jews, some on the Lower East Side, walking distance from Ansche Chesed on Norfolk Street, the Orthodox congregation founded by German immigrants in 1828. In the same urban population, however, were Jewish cooks with a radically new outlook on the dietary laws: women serving oyster stew, baked ham, and creamed chicken casserole, a full menu of forbidden foods.

The readiness to break away from tradition, more pronounced among Germans than any other Jewish group, had its roots in a wider cultural movement that began in Germany in the late eighteenth century, the Haskalah, or Jewish Enlightenment. For centuries, German Jews had lived as outsiders on the distant fringes of the wider Christian society. Their communities were self-governing and inward-looking, sealed off from the culture around them. But not entirely. Inspired by the European Enlightenment, eighteenth-century Jews began to question their own separateness. Men like Moses Mendelsohn, the eighteenth-century German scholar and philosopher, argued for the value of a secular education for Jewish school kids, a revolutionary suggestion at a time when education meant one thing only: the study of Torah. Mendelsohn's ideas took hold, so by the middle of the nineteenth century Jewish children were learning to read and write in German, their passport to the larger world of secular thought. German Jews discovered Goethe and Schiller, along with German translations of the European classics. Even in the countryside, far from the intellectual ferment of the big cities, Jewish families spent the evenings reading aloud from Voltaire and Shakespeare.

The new openness in education encouraged other forms of change, particularly in the cities. Here, "modern-thinking" Jews developed a more relaxed approach to religious observance. On the Sabbath, shopkeepers kept their stores open. Men shaved their beards; women abandoned their traditional bonnets, trading them in for wigs, or went bareheaded. In their synagogues, modern-thinking Jews rejected traditional forms of worship, installing organs and choirs, the rabbi standing before his congregants in a long black frock and delivering sermons, very much like his Christian counterparts.

In the German-Jewish kitchen, a quiet revolution was likewise in progress. For the first time, home cooks felt they could choose among the food laws, holding on to some, dropping others. Of course, the willingness to improvise fell along a sliding scale, with each cook determining her own culinary threshold. Some abandoned the time-consuming practice of salting and soaking their meat, the traditional method for drawing out blood, a substance banned from the Jewish table. Others gave up on kosher meat entirely and started shopping from the gentile butcher. In private recipe collections, we see Jewish cooks experimenting with pork and other forbidden foods. Outside the home, Jews began patronizing Christian-owned establishments, while Jewish-owned eateries, like the palatial Restaurant Kempinski in Berlin, served oysters, crayfish, and lobsters.

Changing food attitudes were carried to America, setting the stage for a remarkable document. *Aunt Babette's Cook Book*, published in 1889, was intended for young Jewish cooks, but only the most "modern thinking" or, as we might say, assimilated. The real-life "Aunt Babette," a home-maker named Bertha Kramer, lived on Chicago's Prairie Avenue, the city's ritziest address. Her neighbors included the Pullmans, the McCormicks, and the Fields of Marshall Fields, along with a sprinkling of affluent Jews, most of German descent. At age thirty-six, Kramer began a recipe column for *The American Israelite*, a weekly Jewish paper with a national readership. It was here, writing under a pseudonym, that she developed her distinctive writer's persona, the cosmopolitan "Babette."[8]

With a Star of David emblazoned on its title page, *Aunt Babette's Cook Book* was published by the Bloch Printing Company, America's premier Jewish publisher. Despite its Jewish pedigree, the recipes in *Aunt Babette's*

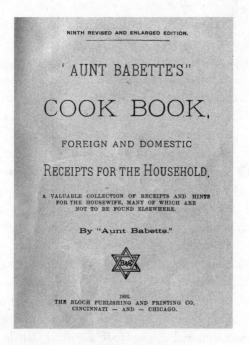

Title page for "Aunt Babette's" Cook Book.

systematically, brazenly, play havoc with the food commandments, including the prohibitions against pork, shellfish, wild game, blood, and mixing meat and dairy. Shrimp and lobster salads, oysters on the half-shell, chopped ham mixed with cream, broiled squirrel, venison, and rabbit pie—all ritually forbidden foods—find their way into *Aunt Babette's*, but so do recipes for matzoh pudding, Purim doughnuts, and gefilte fish, resulting in an eclectic feast of old and new, foreign and domestic, all in one volume. The freewheeling approach continues beyond the recipes into the "valuable hints" section. Here, at the back of the book, the reader will find a home remedy for a sore throat that involves swathing the patient's neck in raw strips of bacon.[9]

Babette's position on kashruth is very much in tune with the mindset of her time and place. In separating the pure (kosher) from the impure (*treyf*), she defers to contemporary standards of good hygiene over ancient law, divine or otherwise. She sums it up this way: "Nothing is trefa that is healthy and clean," dispensing with five thousand years of culinary tradition in a few well-chosen words. Living on Prairie Avenue, Kramer worshipped at the Sinai Temple, the most reformed of any Reform congregation in the United States. Sinai offered its members church-style family pews, a choir, and a service conducted in English, not Hebrew. This was all fairly standard in Reform circles. Sinai went a step further, however, when it moved the Jewish Sabbath to Sunday. As a rule, the Jews who belonged to Sinai were socially ambitious types, eager to hobnob with their Gentile neighbors. Many, including Bertha and her husband, were listed in Chicago's *Blue Book*, a directory of the city's "most prominent householders." Most, by far, were church-going Christians. *Aunt Babette's Cook Book* captures the food attitudes of this particular milieu, immigrant Jews experimenting with a new kind of Judaism and a new culture—social, political, and culinary.

In New York, the first wave of German-Jewish immigrants, most from Bavaria, latched on to the city's older and well-established community of Sephardic Jews. In the 1830s, they joined the Sephardic congregation, Shearith Israel, founded in the seventeenth century, when New York was still New Amsterdam, and married into Sephardic families, blending into the city's Jewish aristocracy to the best of their ability. Tensions developed, however, as the number of Germans continued to swell. The decisive break came in the mid-nineteenth century, as the first generation of modern-thinking Germans landed in New York, forming their own settlement in the old tenth ward of the Lower East Side, then part of *Kleindeutschland*. Here, in a converted church on Chrystie Street, they established New York's first Reform synagogue, Temple Emanuel. With their assimilated habits and enlightenment ideals, the new immigrants were philosophically out of step with New York's Jewish elite. The Sephardim (along with their Bavarian brothers and sisters) were Orthodox, with a strong sense of their own exalted history, but the newly arrived German Jews had reasons to feel superior too. They were cultured, educated, some with university degrees, and solidly middle-class.

In the realm of food, they took full advantage of enlightenment principles to share in the gastronomic culture of their adopted country. The following account shows the free reinterpretation of ancient custom as it played itself out in the Jewish dining room. The meal described below took place in New York shortly after the Civil War:

> A friend of mine, not long since was invited to dine with a wealthy Jew whose name is well known among the most eminent businessmen of the city. The table was elegantly spread, and among the dishes was a fine ham and some oysters, both forbidden by the law of Moses. A little surprised to see these prohibited dishes on the table and anxious to now hear how a Jew would explain the introduction of such forbidden food, in consistency with his allegiance to the Mosaic law, my friend called the attention of the Jew to their presence. "Well," said the host, "I belong to that portion of the people of Israel who are changing the customs of our fathers to conform to the times and country in which we live. We make a distinction between what is moral in the law, and, of course, binding, and what is sanitary. The pork of Palestine was diseased and unwholesome. It was not fit to be eaten, and therefore was prohibited. But Moses never tasted a slice of Cincinnati ham. Had he done so, he would have commanded it to be eaten. The oysters of Palestine were coppery and poisonous. Had the great law-

*giver enjoyed a fry or stew of Saddlerocks or Chesapeake Bay oysters,
he would have made an exception in their favor."* [10]

When this dinner was held, circa 1869, the presence of ham or shellfish
on a Jewish table required explaining. By the end of the century, the same
menu was unremarkable, standard fare among the city's Reformed Jews.
Or, rather, it was almost standard. Still, even the most assimilated Jew
felt a certain unshakable reticence with regard to pork. Shellfish, however, was another story entirely. Here was a food thoroughly embraced
by assimilated eaters, in their homes and in public, too. Most surprising,
perhaps, shellfish were commonly served at Jewish-sponsored events, like
the infamous "trefa banquet" held in Cincinnati in 1883, to celebrate the
first ordination of rabbis from the Hebrew Union College. On the menu
that evening were Little Neck clams on the half-shell, soft-shell crabs
l'Amérique, and finally, "Salade of Shrimp."

In private kitchens, the Jewish home-cook experimented freely with the
whole array of crustacea. *The Famous Cook Book*, another synagogue-based
charity cookbook, this one from Seattle and published in 1908, gives five
separate recipes for clam chowder, each one contributed by a different
member of the congregation. Clearly, the dish was a regional favorite growing out of the local clam trade. A few pages further on are recipes using
Dungeness crab, another creature indigenous to Pacific waters. *The Famous
Cook Book* is just one example of a much broader trend. In fact, few Jewish
charity cookbooks do not include at least a handful of shellfish dishes.

The assimilated cook made full use of the shellfish available to her,
depending on the local markets. By the century's end, however, one
marine animal was available just about everywhere. It was the oyster, the
Jews' favorite forbidden food. This appreciation was a reflection of the
larger American culinary culture. Throughout the nineteenth century,
oysters were consumed with equal gusto by society swells and poor working stiffs, men and women, East Coasters and West Coasters. Perhaps no
other food held such universal appeal. By the 1870s, New York alone
was home to 850 oyster eateries, some grandly decorated in true Gilded
Age style, others no more than a stall at the market. And thanks to the
newly constructed railroad, the oyster craze penetrated to the middle of

the country as well. For the assimilated Jew, it was impossible to resist the tug of the oyster—more so, it seems, than other *treyf* foods. Rabbi Wise, whom we have already met, personally refrained from everything *treyf* with the exception of oysters, which he claimed were technically kosher, their shells equivalent to the scales of a fish, protecting the bivalve from "poisonous gases in the water." In a similar spirit, one Kansas City rabbi argued that oysters were not, in fact, shellfish at all, but rather a form of underwater plant. These were the kinds of legal defenses put forth by clergymen, but down in the culinary trenches, Jewish eaters followed their own logic. More persuasive than any technical loophole was the oyster itself, plump, briny, undeniably delicious. American Jews were incredulous that any food so patently good could be forbidden.

True to her time, Aunt Babette was enamored of the oyster, providing instructions for an all-oyster supper: "In giving an oyster supper always serve raw oysters first, then stewed, fried and so on. Serve nice, white crisp celery, olives, lemons, good catsup, cold slaw and pickles and do not forget to have two or three kinds of crackers on the table." Jewish cooks in New York were equally enthralled. In 1909, the Hebrew Sheltering Guardian Society, a prominent New York charity, published a cookbook. With recipes contributed by society members, *The Auxiliary Cook Book* provides a direct and focused picture of Jewish food ways at a particular moment in time. Among the half dozen oyster recipes included in this slender volume are fricassee of oysters cooked in brown butter, oyster stew, and oyster cocktail. These were the kind of standard American oyster dishes adopted by the assimilated homemaker. But the Jewish cook could be more creative as well, fusing Old and New World culinary traditions in highly unorthodox ways. Perhaps the best example of this is the "oyster noodle kugel," a recipe found in the *Council Cook Book* published by the Council of Jewish Women in 1909: "Into a pudding dish put layers of broad boiled noodles, alternating with layers of oysters dipped in cracker crumbs, with plenty of butter and salt to taste; pour over the whole pint of pastry cream and the juice of the oysters; bake until brown—about twenty minutes."[11]

Like most East Siders, when Natalie Gumpertz departed this world, she left very little behind in the way of documentation. No diary, no book of household accounts, no correspondence, and no family recipes—the kind of detailed and personal records left by middle-class women. The poor, meanwhile, left behind a different class of evidence: census and draft records, marriage licenses, birth and death certificates—the kinds of documents that fill our municipal archives, the city's official memory. According to her death certificate, Natalie Reinsberg Gumpertz was born in Ortelsburg, East Prussia, an out-of-the-way market town with a small community of Jewish merchants and tradesmen. As a girl in Ortelsburg, Natalie would have received a state-sponsored education, in German, but may have spoken Yiddish as well. The region surrounding Ortelsburg was known as the poorest in Prussia, its Jewish population largely impoverished. Natalie's father, however, was a person of at least some means, contributing money to build the town's first synagogue.

Natalie Gumpertz, circa 1890. Mrs. Gumpertz lived on Orchard Street from 1870 to 1886.

Small-town Jews like the Reinsbergs lived on a cultural frontier between Germany's larger, more modernized Jewish communities and the traditional shtetl Jews of Poland and Lithuania. They spoke German and embraced German culture with the same pride of ownership as their Christian neighbors. In their daily lives, they moved with relative freedom among their non-Jewish countrymen, doing business, even socializing with Christians, a pattern rarely found among the Jews of small-town Bavaria. Their version of Judaism, however, was closer to the religion of their Eastern brothers and sisters. They followed Polish religious custom, prayed in an Orthodox synagogue, and led Orthodox lives. Tucked away in the Prussian lake region, they lived beyond the reach of Reform thinking, a movement centered in the cities.

The Reinsbergs also lived on a culinary frontier. Located on the edge of German-speaking Europe, East Prussia felt the culinary influences of Central and Eastern Europe in near equal measure, blending German specialties like wursts and kuchens with typically Slavic foods like borscht and pirogi. Add to this complex scenario the fact that most Prussian Jews were Polish transplants. Reversing the normal eastward flow of Jewish migrants, they carried Polish cooking traditions back into Germany, foods like gefilte fish and cholent, or Sabbath stew. Following the usual Jewish pattern, they held on to familiar foods while adopting local staples. For Prussian Jews, that included dairy products like *schmant* (sour cream) and *glumse* (farmer's cheese), two key foods in the local diet, an abundance of freshwater fish, and Prussian firewater, a combination of grain alcohol and honey. In short, East Prussian Jews were culturally assimilated but traditional in their religious practice. In their kitchens, Prussian cooks followed the food laws as best they could—not always easy in remote towns where the closest source of kosher meat could be miles away.

In the New World, Jewish culinary custom came up against a new set of realities. Living on Orchard Street, Natalie Gumpertz faced a highly precarious financial future. Between 1870 and 1874, her husband, Julius, bounced from one job to another, working as a clerk and then a salesman before returning to his old standby as a shoemaker. Unhinged by the strain of supporting his family, he eventually buckled completely. On October 7, 1874, Julius left the Orchard Street apartment for work and never returned, thus joining the ranks of the East Side's many "missing husbands." (The phenomenon was so common that the *Daily Forward*, the city's leading Yiddish newspaper, ran a regular photo feature called the "Gallery of Missing Husbands.") Natalie tried to find her missing spouse, but he was never located, leaving the young mother to fend for herself.

Abandonment was a special class of hardship reserved for East Side women. All immigrants, however, faced the challenge of plain economic survival. In America, land of the mighty dollar, businesses ran six days a week, pausing on Sundays so workers might catch their breath. Back in Europe, traditional Jews abstained from toil on the Sabbath. In America, the economic pressures to work on Saturday were hard to resist, and many Jews relented. Similar concessions were made in the realm of food. For some, that meant eating kosher at home but sharing in the wider culinary culture when out in public. Others stayed loyal to traditional foods but consumed them in ways unimaginable to the strictly devout. Vivid examples of how this compromise worked appear in the fiction of Fannie Hurst, a German-Jewish writer with a tremendous following in the 1920s. The only daughter of two assimilated Bavarian Jews, Fannie Hurst was born on her grandmother's Ohio farm in 1885 and raised in St. Louis, Missouri. In 1909, she moved to New York to launch her career, and by the 1930s had produced hundreds of short stories and a string of novels, some of which were made into movies (the best known is the twice-filmed *Imitation of Life*).

The characters Hurst was most drawn to were poor urban women— shopgirls, streetwalkers, maids, and tenement housewives, many of them Jewish immigrants. Often, Hurst's stories begin in the tenements, following her characters as they break into the middle class. For the typical Hurst heroine, it's a bittersweet victory, the material rewards of her new middle-class life offset by the demands of assimilation. The fictional Turkletaub family of West 120th Street have made precisely that journey. In the following scene, they have just sat down to an abundant midday meal:

At one o'clock there was dinner, that immemorial Sunday meal of roast chicken with its supplicating legs up off the platter; dressing to be gouged out, sweet potatoes in amber icing; a master stroke of Mrs. Turkletaub's called "matzos klose," balls of unleavened bread, sizzling, even as she served them, in a hot butter bath and light-brown onions; a stuffed goose neck, bursting of flavor, cheese pie twice the depth of the fork that cut in; coffee in large cups.

Just a few hours later, Mrs. Turkletaub is back in the kitchen, preparing Sunday night supper, the second feast of the day:

A platter of ruddy sliced tongue; one of noonday remnants of cold chicken, ovals of liverwurst; a mound of potato salad crisscrossed with strips of pimento; a china basket of stuffed dates, all kissed with sugar; half of an enormously thick cheese cake; two uncovered apple pies; a stack of delicious curlicues, known as "schneken," pickles with a fern of dill across them.[12]

Though still bound to some of the old traditions, the Turkletaubs are willing to defy the religious injunction of mixing meat and dairy. Even more, the family has abandoned the traditional Friday night meal and moved it to Sunday, celebrating the Christian day of rest, but doing it with Jewish food! (Incidentally, the Turkletaubs were not alone. As far back as 1859, a contingent of German-Jewish New Yorkers, many of them bankers, experimented with rescheduling the Sabbath—their attempt to reconcile Judaism with the American workweek.)

The Jewish Sunday dinner is one example of the push-pull relationship that soon developed between immigrants like the Turkletaubs, Jews from traditional backgrounds, and their ancestral foods. Sabbath dishes like chicken soup, brisket, or challah became Sunday dinners in America, suddenly free of the old religious restraints. Holiday foods reserved for special times of the year were drained of their symbolism, and eaten on a whim, whatever the season. In Europe, potato pancakes were a wintertime delicacy inextricably connected with the celebration of Hanukkah.

In New York, they were sold year-round by East Side street vendors. Blintzes in Europe were a symbolic food of spring, traditionally eaten on Shavuot. On this side of the Atlantic, they were standard fare in the scores of East Side dairy cafés.

Newly flush with American dollars, middle-class Jews like the Turkle-taubs turned their dining-room tables into edible landscapes. Three kinds of meats, dumplings and salads, cakes and pie, followed by coffee, then raisins and nuts (a favorite Jewish snack), all in a single meal. Largesse on such a grand scale was impossible back in the Old Country. At mealtime, they unloaded their well-stocked pantries, as if to say: "Look what we have! Incredible, no?"

Downtown in the tenements, where the typical pantry was nine-tenths empty, German Jews relied on a long culinary tradition of "making do." Mainstays of the tenement cook were starchy, cheap, and filling. She was, for example, an expert noodle maker, her kitchen equipped with a

Shoppers inspecting the merchandise on Hester Street, 1898.

"noodle board" for rolling and cutting dough. If her husband was handy, the noodle board was attached to the wall by a hinge, so it folded up neatly when not in use. To slash the cost of homemade noodles, a bargain food to begin with, she bought broken eggs, a common commodity among East Side vendors. Ever prepared, the East Sider carried an old tin cup or a beer mug in her shopping basket. When she came to the egg stand, she cracked the already-shattered eggs into the cup, sniffed them, and eyed them. If they met with her approval, she carried them home, fully cracked, at the vendor's discounted price. All she needed now was a few cups of flour to create a meal.

Chicken noodle soup was a luxury food by downtown standards, and commonly reserved for Fridays or holidays. A midweek meal among German Jews might be pea soup with spaetzle, pebble-shaped nuggets of dough grated straight into the soup pot. Another option was noodles stewed with fried onion and perhaps a piece of crumbled liver or "soup meat." There were also noodles and fruit—dried pears, apricots, or prunes—stewed together over a slow flame. The end result was a stovetop version of noodle kugel, sweet and satisfying.

Where Bavarian cooks looked to noodles as their starch of choice, Jews from Prussia, Posen, and all points east relied on potatoes. No people on earth could equal the Irish in potato consumption, but the Jews of Eastern Europe came close, remarkable when you consider the potatoes' relatively late arrival on the Jewish food scene. Actually, widespread potato cultivation came surprisingly late to Europe in general. For a solid two hundred years after its European debut in the sixteenth century, the potato remained an obscure sample of New World flora. It was studied by botanists and grown as an ornament, but rarely eaten. The one place it made any headway as a food was the court kitchen, served now and then as a novelty item to aristocratic dinners. Commoners were less favorably impressed. Most Europeans feared the potato as a carrier of disease or scorned it as a food of heathens. Those who tried it, found it plain unappetizing, better suited to livestock than people. The potato showed its virtues mainly when other crops failed. It could withstand long stretches of cold temperatures better than most grains, and matured more quickly. (It takes two to three months for potatoes to mature, versus ten months for wheat.) During times of political turbulence, the edible part of the potato plant stayed buried in the earth, safe from marauders and their torches.

By the start of the eighteenth century, extensive potato farming was found throughout the Low Countries and in parts of France and western Germany. By the middle of the century, Frederick the Great, king of Prussia, had grasped the potatoes' usefulness as an insurance crop—a backup food when other foods were scarce—and in 1744 commanded wide-scale potato planting. The order took roughly thirty years to implement, but in the end Frederick prevailed through a mix of force, coaxing, and education. (The king distributed free seed potatoes, along with an instructional handbook, to farmers.) By the end of his reign, his dream was realized: Prussia was rich in potatoes. Frederick's campaign introduced potatoes to the Prussian Jews, numbering over a hundred thousand and a very receptive audience. Even more, he brought the potato to the edge of Eastern Europe and its vast Jewish community. It was just a matter of decades before it spread across Poland and eventually Russia.

Jews embraced the potato for many of the same reasons as the Irish. It was a high-yielding, fast-growing plant, even in poor soil, while the potato itself was dense in both calories and nutrients. Like the Irish, the Jews cooked their potatoes whole, either boiled or roasted, and peeled them at the table. And, like the Irish, Jews "kitchened" their potatoes with some form of dairy food, perhaps a glass of buttermilk, or maybe a few bites of protein, usually herring. When the herring ran out, they dipped their potato into the pickling brine. But sometimes there was nothing—no fish, no brine—and the work of seasoning the potatoes was left to the eater's imagination. The following domestic scene comes from the Yiddish story "When Does Mame Eat?" by Avrom Reisen. It highlights nicely the power of suggestion as a culinary tool:

> In the morning, when Leybele got up, he saw Mame standing by the oven, sticking one pot in and sliding the other out—she was so handy!

And she had forks of all kinds—a big one for a big pot, a small one with a small pot, and the middle-sized tines to go with the middle-sized pot. Then, when she removed one pot, she checked its contents, blew away the foam, and put it back in the oven.

What did she cook in those pots? One of them, a day's worth, was dedicated to hot water, nothing else. "A house," said Mame, "has to have hot water." The second pot was surely filled with potatoes, called "fish potatoes" when cut up. But there wasn't any fish . . . They didn't eat fish during the week. "Fish," said Mame, "is very expensive." And fish potatoes, even without fish, were tasty too. In any case, Mame was a skilled housewife, that he knew for sure.[13]

Boiled potatoes with imaginary fish. This was Jewish potato cookery at its most austere. Alongside it, however, Jewish cooks developed a full repertoire of more elaborate potato-based dishes: puddings, dumplings, breads, soups, even baked goods.

Potato kugel, a Sabbath mainstay of Polish Jews, was made from grated raw potatoes and onions, goose fat, eggs, and bread crumbs, all mixed together and baked overnight until golden with crisp, brown edges. *Golkes*, another Polish specialty, were chewy potato dumplings. They were made from grated raw potatoes that had been wrung dry in a towel, then mixed with flour and eggs, rolled into balls, and boiled. *Golkes* were typically eaten in soup or maybe in a bowl of hot milk. (Potatoes and milk, or some form of dairy food, were a very common pairing in the Jewish kitchen, just as in Ireland.) A Lithuanian food called *bondes* was made from grated potatoes mixed with rye or buckwheat flour to form a rough dough. After a good kneading, the dough was placed on cabbage leaves and baked. Lithuanian mothers gave *bondes* to their children to bring to school, but a gourmet treat was *bondes* hot from the oven with sour cream. Another Lithuanian creation, known as "mock fish" (relative of "fish potatoes"), was made from sliced potatoes cooked with onion and goose fat. In the summer, when the cows were producing, "mock fish" was converted into a dairy dish, the potatoes and onions cooked with butter and sour cream. Potatoes, onion, and fat—the Jewish cook explored every conceivable permutation of these three core ingredients, the more fat the fancier the dish. The most extravagant of all was latkes, potato pancakes fried in sizzling pools of precious goose fat.

Many of the potato dishes made by German Jews were crossover foods, their origins in the Gentile kitchen. The potato dumpling was one of them. Early immigrant cookbooks offer a narrow sampling of the thousands of variations that must have existed, each cook playing with the basic dumpling formula to match her tastes and budget. Dumpling dough was typically made from cooked potatoes, egg, and just enough flour to form a workable but not-too-stiff dough. The dough was rolled into balls and boiled, then baked or lightly fried. A typical dumpling filling was bread crumbs or cubed bread, both fried with onion. Aunt Babette gives a recipe for "Wiener" potato dumplings, using these standard elements, only her version is formed like a jellyroll that is cut into segments so the dumplings are sausage-shaped. Once boiled, the Wiener dumpling is bathed in onion-scented goose fat. As accompaniments, she recommends "sauerkraut, sauerbraten, or compote of any kind." If the dumpling itself was a mild-flavored food, the Jewish cook surrounded it with strong tastes—savory, sweet, aromatic, and sour—sometimes all on one plate. Jewish cooking today has a reputation for blandness, not entirely unearned. A hundred years ago, however, the label would have never stuck. The nineteenth-century Jewish cook specialized in bold flavors and complex flavor combinations, sweet with sour being a particular favorite. As a result, native-born Americans often looked down on Jewish cuisine as "too highly seasoned," which in their eyes was both unhealthy and uncouth.

Noodles and potatoes were largely interchangeable in the Jewish kitchen, receiving many of the same treatments. Both foods, for example, could be savory or sweet, cooked with liver and onion on one hand, or sugar and cinnamon on the other. Like noodles, potatoes were sometimes paired with fruit. *The Famous Cook Book* gives a recipe for potato puffs (the dumplings' new American name) stuffed with cooked prunes. Here, the boiled dumplings are dotted with butter, baked, and "served as a vegetable," or, as the author suggests, sprinkled with cinnamon and sugar to make a dessert. In middle-class kitchens, dumplings were reduced to the status of side dish. In the tenements, potato dumplings with sauerkraut or fruit-filled dumplings with sour cream made a cheap and flavorful midweek supper.

As immigrant Jews moved up and out of the tenements, they took along the old foods of necessity. These were the dishes they had once depended on for survival. Noodles and potatoes were just two of them. Another was stuffed turkey neck, along with many comparable stuffed or filled creations——a time-honored strategy for stretching protein, now eaten just for pleasure. There were also holiday foods like latkes and matzoh brei, which transcended the uptown/downtown divide. A more fundamental link, however, centered on the issue of fat, a traditional preoccupation of the Jewish cook. Aunt Babette, for example, betrays her fat fixation each time she instructs readers to "skim off every particle of fat," from any simmering soup or stew, but not to discard it, as the modern cook would. On this point she is very clear. "I would like to suggest right here," she writes, "never throw away fat." Instead, she tells readers to save it for use as a seasoning, a cooking medium, or a shortening in baked goods. Aunt Babette's instinct for the preciousness of fat cut across distinctions of class, connecting well-heeled immigrants with reformed sensibilities to their working-class cousins. Even more, it transcended regional distinctions, connecting German Jews with their brothers and sisters from Eastern Europe.

The Jewish concern with fat was born in Germany at the dawn of Ashkenazi culture. As we already know, the Ashkenazim were prolific culinary borrowers, adopting many local staples from the German peasantry. One staple, however, was strictly off-limits. For Gentile cooks, the pig was a veritable walking larder. The peasant housewife fed it on kitchen scraps, and it supplied her with hams, sausage, bacon, feet for pickling, and blood to make puddings. Most valuable of all, it supplied her with lard, the peasant's primary cooking fat. Lard, of course, was forbidden to Jews and so was beef tallow. Butter was problematic, because of the prohibition against mixing meat with dairy. Historically, Mediterranean Jews had relied on olive oil, an impractical option in northern climates. So, the

Ashkenazim turned to poultry fat, the food we know today as schmaltz.

In its first incarnation, schmaltz was derived not from chicken but from geese. As early as the eleventh century and possibly before, German Jews had taken up goose farming, raising birds that were stunningly plump, veritable fountains of schmaltz. Their secret was force-feeding. Jewish-raised geese led normal lives until their final weeks. A month or so before slaughter, they were subjected to a rigorous feeding regimen in which compacted pellets of grain or dough were pushed down the animal's throat. As German Jews migrated east, they carried the technique into Poland and Russia, where goose farming developed into a Jewish niche occupation most closely identified with women.

In towns across central and eastern Europe, Jewish women kept two or three, or, in some cases, a small flock of geese. During the summer months, the birds were free to walk the streets, their mistress trailing behind, waving a switch. In late autumn, they were put into a "goose house." Now the force-feeding began. The nineteenth-century German cookbook author Rebekka Wolf gives the following instructions for goose fattening or *"Ganse zu nudelen,"* literally translated as "to dumpling geese":

> Make dough from coarse meal and bran, adding a hand-
> ful of salt, some beechwood ashes (if you have some) and
> water so it forms a good ball in your hand, and make from
> it dumplings a half a short finger long and a fat finger thick
> and then dry them on a hot pan or at the baker. In the
> beginning a goose receives four pieces per serving, which is
> given four times a day or 16 per day. Do this for three or
> four days and then for a few days give seven pieces per serv-
> ing, then nine, then 11 and then at most 13 pieces, where
> you stay until the goose is fat, which is best felt on the bot-
> tom of the bird.[14]

Just before Hanukkah, women brought the geese to the local *shochet* (ritual slaughterer) to give them a proper death. It was the women's own job, however, to disassemble the bird. Geese had to be plucked, salted (to draw out the blood), scalded, then broken down into parts. The breast was smoked, the skin fried to make *gribenes* (the kosher answer to bacon); the neck was stuffed and roasted or stewed, while the wings, feet, and giblets were saved for the soup pot. The feathers were used for bedding. The bird's enlarged liver, the food we know as foie gras, was roasted and dutifully fed to the children as a nutritional supplement, the same way American children were given doses of castor oil. Finally, the fat was rendered and poured into earthenware jars for use throughout the year, or as long as the cook could stretch it.

The Jewish cook used goose fat for frying, baking, braising, enriching, moistening, and seasoning. It was a stock ingredient in her best, most succulent foods. These were the dishes prepared for the holidays, the kugels and cholents and kreplach, to name just a few. Warming, satiny, with a faintly nutty aftertaste, it imbued foods with a pleasing heaviness, a liability, perhaps, to the modern diner, but for the calorie-deprived a virtue. To the Jewish palate, fat represented the essence of goodness. The nineteenth-century Jewish homemaker brought her reliance on geese and all its by-products to the Lower East Side, where she continued her traditional role as a poultry farmer. Amazingly, immigrants raised geese in tenement yards, basements, hallways, and apartments as well, transplanting a rural industry to the heart of urban America.

Tenement goose farms belonged to a well-established tradition of animal husbandry on the Lower East Side. During the first half of the nineteenth century, neighborhood streets served as a communal feeding trough for wandering pigs, a common sight through the 1850s. East Side pigs were the property of poor New Yorkers who had set their animals free to scavenge for food, feasting on refuse until they were ready for slaughter. In life, they acted as street cleaners; in death, they supplied their owners with an abundance of virtually free meat. The majority of New York's pig keepers were recently arrived Irish immigrants, veteran pig farmers from way back. In the years following the Potato Famine, as the number of New York Irish ballooned, so did the number of swine. In 1842, the city was home to roughly ten thousand wandering pigs. Before the decade was up, that figure had doubled.

Until the 1860s, repeated attempts to rein in the pigs were only marginally successful. The work of removing the animals from Lower Manhattan fell to the newly created sanitary police, a specialized unit within the larger police force, which was established to protect the health and safety of New Yorkers during a period of very rapid population growth. The squad's four main areas of responsibility were ferries, factories, slaughterhouses, and, most relevant to our story, tenements. Creation of the sanitary police was the first of several related developments in the campaign for a cleaner, more salubrious New York that unfolded in the 1860s. In 1865, the Citizens' Association, a group of reform-minded New Yorkers, launched a comprehensive, block-by-block, sanitary survey of Manhattan with special focus on conditions in the tenements. The fruit of their labor was the 504-page *Report of the Council on Hygiene and Public Health of the Citizen's Association of New York upon the Sanitary Conditions in the City*. One year later, at the council's strong urging, the city established the Metropolitan Board of Health, America's first permanent public-health agency.

With the pig situation under control, "sanitarians" shifted their energy to a new problem: the tenement poultry farms, which began to spring up in the 1870s in heavily Jewish areas of the Lower East Side. Where urban pigs were on public display, tenement poultry farms posed a more insidious threat, hidden away as they were in the same living space as humans. Sanitary inspectors were aghast. While they strived for a professional tone, the sense of horror in their reports is palpable even over a century later. The following description is from 1879:

> One who has only seen poultry kept in the country, where the only nuisance attributable to them is scratching up seeds, can hardly realize what a terrible nuisance they may cause in the city. Where many fowls are huddled together in contracted quarters, they keep up an incessant clucking and cackling, and the odor that rises from them is overpowering. In New York the Board of Health has carried on a struggle

for some years, with occasional breathing-spells for both combatants, against the practice of keeping poultry for sale in the manner practiced by Polish and Russian Jews. On the plea that their religion requires them to eat only those fowls that have been killed in their sight by a killer authorized under their ritual, they fill the places where they live with chickens, turkeys, ducks and geese. These poor fowls are huddled together in coops, or crowded into pens, generally in the basement of the house, and make an incessant noise. The smell, too, from fifty or a hundred geese is indescribable and intolerable. And these people live in an adjoining room, and wonder that any person finds their practice obnoxious.[15]

Thursdays and Fridays, Jewish shoppers would descend on the farms and select their bird by blowing between the tail feathers for a glimpse of the skin. The yellower the skin, the fatter the bird. With the housewife looking on, the animal was slaughtered in the yard by an itinerant *shochet*, his only equipment a curved knife and a barrel filled with sawdust to collect the blood.

Though East Side farmers trafficked in all types of domestic fowl, their bestseller was geese. In tenements along Bayard, Hester, Essex and Ludlow streets, where basements doubled as goose pens, East Side goose farmers did a booming business despite frequent raids by the sanitary police. Some were issued fines, others were hauled off to jail, but Jewish farmers persisted, just as the Irish had done a generation earlier. The Jewish demand for goose meant steady profits, and East Side farms continued to multiply along with the Jewish population.

In later years, Jewish goose-farming expanded from a cottage industry to a major commercial enterprise, with large poultry yards lining the East River. By the 1920s, the kosher poultry trade was lucrative enough to attract organized crime, and a racketeering operation grew up around the city's kosher slaughterhouses.[16] By 1900, the tenement goose farmers had been reduced to piecework as "dry pickers," or feather-pluckers, paid just a few cents per bird. Jewish women were also hired as "goose stuffers," using skills they had acquired centuries ago and passed down from mother to daughter. A widely printed newspaper story from 1903,

titled "Some Queer East Side Vocations," describes what became of the
birds' yellowy-beige, overgrown livers:

> They are made up into a sort of paste, chopped fine with onions,
> garlic, and other strong-smelling seasoning, or are fried in fat, after
> being dipped in cracker crumbs. . . . When these delicacies are to be had
> on the menu of a kosher restaurant, a card is hung on the window
> to that effect just as a Christian restaurant announces the fact that it
> has soft-shell crabs or North River shad.[17]

As for the fat-laden skin, it was diced and heated to produce "the rich,
thick grease" better known as schmaltz.

If a whole fattened goose was beyond the means of the tenement
homemaker, she could buy odds and ends—giblets, necks, wings, and
skin—cheap but flavorsome parts, if one knew how to handle them. A
savvy cook, for example, could create a faux foie gras using goose fat,
giblets, and regular chicken livers:

IMITATION PATE DE FOIE GRAS

Take as many livers and gizzards of any kind of fowl as you
may have on hand; add to these three tablespoons of chicken
or goose fat, a finely chopped onion, one tablespoon of pungent sauce, and salt and white pepper to taste. Boil the livers
until quite done and drain; when cold, rub into a smooth
paste. Take some of the fat and chopped onion and simmer
together slowly for ten minutes. Strain through a thin muslin
bag, pressing the bag tightly, turn into a bowl and mix with
the seasoning; work all together for a long time, then grease a
bowl or cups and press this mixture into them; when soft cut
up the gizzards into bits and lay between the mixture. You
may season this highly, or to suit taste.[18]

In the tenement kitchen, the luxuriousness of goose fat elevated the most prosaic ingredients. In the following recipe, a dab of goose fat transforms onion and rye bread into a delicacy.

Lightly sauté one yellow onion, thinly sliced, in four table-spoons goose or chicken fat. Spread cooked onion on good rye bread. Season generously with crushed black pepper. For a more substantial snack, top with sliced hardboiled egg.[19]

As modern methods of chicken breeding improved in the twentieth century, the goose lost its place of prominence on the Jewish table, replaced by its smaller, more economical cousin.

An East Side "chicken market," 1939. By the 1920s, chicken had largely replaced geese in the Jewish immigrant's diet.

Chicken fat now took the place of goose fat as the Jew's favorite cooking fat. The raison d'être for poultry fat of any kind, however, was essentially erased by the invention of scientifically engineered cooking fat derived from vegetables. The new hydrogenated fats came with many names; Flake White, Spry, Snowdrift, and Nyafat, a vegetable shortening pre-seasoned with onion, are just a few. The name best known today, however, is Crisco, a product created by Procter & Gamble, a Cincinnati-based soap manufacturer, in the years before World War I.

Tellingly, Crisco was originally developed as a cheaper alternative to the lard and beef tallow traditionally used in soaps and candles. Looking to expand its uses, Procter & Gamble introduced Crisco to American cooks in 1911, presenting it as a more economical and "more digestible" substitute for lard and butter. There was nothing Jewish about Procter & Gamble, but in time the company recognized the value of its product to Jewish cooks. Most important, Crisco was pareve, or neutral—a ritually permissible partner to either dairy or meat. The Jewish cook could bake with it, braise, or fry, pairing it with any ingredient she chose. No other fat in the Jewish pantry was as versatile, not even the beloved goose schmaltz. *Crisco Recipes for the Jewish Housewife*, a promotional cookbook published by Procter & Gamble in 1933, allowed the kosher cook to imagine the freedom awaiting her in the blue-and-white can. Clearly aimed at immigrants (the book was published with both Yiddish and English recipes), it represents the demise of poultry fat as a Jewish staple, bringing to a close a millennium of culinary tradition.

For a brief time in the 1870s, 97 Orchard was home to a mix of Irish, German, and Jewish families. For each of these groups, dinnertime likely included potatoes, herring, onion, cabbage (either pickled or fresh), lard or goose fat, and some form of dairy. These were the foods that sustained the nineteenth-century East Sider, regardless of national background. Despite this facade of a common cuisine, however, each immigrant group brought to the dinner table food-based assumptions that shaped their experience of eating. German East Siders held up their ancestral foods as cultural trophies, celebrating their German past in grand-scale and very public eating events. The Irish, by contrast, celebrated with drink and music and dance, but confined the serious work of feeding themselves to the privacy of their homes. As for the Jews, they came to the dinner table with a distinct and highly developed zest for eating, a sensibility so evolved and pronounced it deserves a term of its own: food-joy. Like the fondness for fat, Jewish food-joy was born of scarcity. (A firsthand knowledge of hunger was perhaps the single greatest common denominator among all East Side immigrants.) But it was more than that, too. Jewish food-joy was grounded in the elaborate system of culinary laws and rituals that transformed the everyday business of eating into a sacred act. As the rabbis explained, God had honored the Jews with a culinary mandate. Where Gentiles could eat as they pleased, Jews were given the dietary laws as an outward sign of their special relationship with God. In return, they obeyed the laws as a show of devotion, turning mealtime into a form of sacrament. For Jews who followed the letter of the law, blessings were required for every morsel that crossed their lips, continuing reminders of food's divine provenance.

A core belief in the sacredness of food was the linchpin of Jewish food culture, always simmering in the Jewish eater's thoughts. Fridays, in the Jewish kitchen, cooks like Mrs. Gumpertz saved their best ingredients and marshaled their skills for a meal of cosmic significance: Sabbath dinner, a celebration of nothing less than the miracle of creation. The midweek chanting now exploded into full-throated singing and the Jews feasted, even if that meant living on tea and potatoes for the rest of the week. Even for the poorest Jew, Sabbath dinner was a meal set aside for enjoying, quite literally, the sacred fruits of creation. Skimping was out of the question. At the Sabbath table, Jews recast the earthly pleasure of eating as a show of gratitude to the heavenly creator. Pleasure, in fact, was mandatory. After all, God's first commandment to Adam and Eve was to savor the bounty of Eden.

On the Lower East Side, the pleasures of food became a common theme in immigrant writing. In the fictional world of Anzia Yezierska, a Russian-born writer who immigrated to New York around 1890, food was the proverbial ray of light in an otherwise bleak experience. The typical Yezierska heroine is the young East Side woman, oppressed by the ugliness of the ghetto, exploited by her sweatshop boss, but still bursting with life. Craving beauty, she finds it in food. The following exchange between the despondent Hannah Brieneh and her neighbor, Mrs. Pelz, is from *Hungry Hearts*, Yezierska's first collection:

> *"I know what is with you the matter," said Mrs. Pelz. "You didn't eat yet today. When it is empty in the stomach, the whole world looks black. Come, only let me give you something good to taste in the mouth. That will freshen you up." Mrs. Pelz went to the cupboard and brought out the saucepan of gefulte fish that she had cooked for dinner, and placed it in front of Hannah Brieneh. "Give a taste my fish," she said, taking one slice on a spoon, and handing it to Hannah Brieneh with a piece of bread.*
>
> *"Oy wei. How it meltz through all the bones," she exclaimed, brightening as she ate. "May it be for good luck to all," she exalted, waving aloft the last precious bite. Mrs. Pelz was so flattered that she even ladled up a spoonful of gravy.*
>
> *"There is a bit of onion and carrot in it," she said, as she handed it to her neighbor.*
>
> *Hannah Brieneh sipped the gravy drop by drop, like a connoisseur sipping wine.*
>
> *"Ahh. A taste of that gravy lifts me to heaven!"*[20]

Such is the magic of food-joy! If God created the fruits of the earth, a second act of creation took place in the kitchen, where homemakers performed their most valued task—cooking for the family. Appreciative Jewish eaters expressed their gratitude with extravagant praise, an echo to the food blessings, only these words were for mortal ears.

As tenement Jews moved up in the world, they became proficient in English; they changed their manner of dress and often their names, and adopted new habits. Men took up cigars; women coiffed their hair and scented their handkerchiefs. Jews who made the voyage to Upper Manhattan (anywhere above 14th Street) invented a hybrid culture that was reflected with particular clarity in the way they ate. One early chronicler of that culture is the largely forgotten writer Henry Harland, also known as Sidney Luska, Harland's nom de plume in the early part of his career. Beginning in the 1880s, the Catholic-born Harland went undercover as a German Jew to write a series of romantic novels set mainly on the Lower East Side. The most successful was *Yoke of the Thorah*, about a young East Side Jew who marries—tragically, as it turns out—into an uptown family of shirtwaist magnates.

Sunday dinner in the Blums' Lexington Avenue townhouse is a patchwork of seemingly incompatible foods and food traditions somehow pieced together in a way that makes sense to those at the table. The meal, which begins with the traditional blessing, fills the entire afternoon: ten courses and five kinds of wine, followed by an after-dinner liqueur and cigars for the men; in other words, the quintessential Gilded Age banquet. The banqueters, however, eat with the same earthy sense of relish as their downtown brothers and sisters:

> *During the soup, not a word was spoken. Everybody devoted himself religiously to his spoon. At last, however, leaning back in his chair, heaving a long-drawn sigh, and wiping the tears of enjoyment from his eyes, Mr. Blum exclaimed fervently, "Ach! Dot was splendid soup!" And his spouse wagged her jolly old head approvingly at him, from across the table, and gurgled: "Du lieber Gott!"*
>
> *This was the signal for a general loosening of tongues. A very loud and animated conversation at once broke forth from all directions. It was carried on, for the most part, in something like English; but every now and then it betrayed a tendency to lapse into German.*
>
> *"Vail," announced Mr. Blum, with a pathetically reflective air, "when I look around this table and see all these smiling faces, and smell dot cooking and drink dot wine—my Gott!—dot reminds me of the day I landed at the Baittery forty-five years ago, with just exactly six dollars in my pocket. I didn't much think then I'd be here today. Hey, Rebecca?"*

"Ach, God is goot," Mrs. Blum responded, lifting her hand and casting her eyes toward the ceiling.[21]

Rich as Midas, but still tears of enjoyment over a bowl of soup. The soup recipe below is from Aunt Babette:

White Bean Soup

To one quart of small dried beans add as much water as you wish to have soup. You may add any cold scraps of roast beef, mutton, poultry, veal or meat sauce that you may happen to have. Boil until the beans are very soft. You may test them in this way: Take up a few in a spoon and blow on them very hard, if the skin separates from the beans you may press them through a sieve, or take up the meat or scraps and vegetables and serve without straining. Add salt and pepper to taste. A great many prefer this soup unstrained. The water in which has been boiled a smoked tongue may be used for this soup. This may be thickened like split pea soup. Excellent.[22]

Lentil soup was another hearty staple of the German-Jewish homemaker—thick like a stew and smoky-flavored from the addition of sausage. The recipe for lentil soup below comes to us from Kela Nussbaum, a Bavarian homemaker born to a long line of accomplished home cooks, who immigrated to the United States shortly after World War II. Many of Mrs. Nussbaum's recipes, including this one, were kept in the family for centuries, preserved and passed down in handwritten form. Mrs. Nussbaum was the great-granddaughter of Rabbi Bamberger of Wurstberg, the illustrious nineteenth-century educator. In accordance with family tradition, lentil soup was known as "hiding soup" in the Nussbaums' kitchen, a reference to the way the sausage tended to "hide" amid the lentils.

Lentil Soup

1 1-pound bag brown lentil
1 tablespoon vegetable oil
1 large onion, finely chopped
3 stalks celery, finely sliced
2 cloves garlic, minced
1 ringwurst (approximately 1 pound)
2 tablespoons flour
salt and pepper

Soak lentils in abundant cold water until they expand, about 2 hours. Drain and set aside. In a large soup pot, sauté the onion and celery until soft and onion turns pale gold. Add garlic and cook until fragrant. Add ringwurst, whole, drained lentils, and 7 cups of water. Bring to a gentle boil. Turn down heat and simmer until lentils are barely tender. In a cup, mix flour with a few tablespoons of cooking broth to form a roux. When free of lumps, return roux to the soup pot. Stir and continue cooking until lentils are fully tender but still hold their shape. Remove ringwurst, slice into discs, and return to the pot. Season with salt and pepper.[23]

Natalie Gumpertz resided on Orchard Street for a total of fifteen years, four years with her husband, and eleven years without him. But while she remained stationary, the world around her was in motion. German East Siders, most of them Protestant, were leaving the neighborhood, making room for the great influx of Russian Jews, which began in the early 1880s and continued for another twenty-five years. With this shift, the language of the street switched from German to Yiddish, followed by the shop signs. In 1886, John Schneider closed his basement saloon after nearly a quarter-century. Shortly thereafter, the space was taken over by two Jewish merchants, Israel Luftgarden, a butcher, and Wolf Rodensky, who operated a grocery. Both men lived in the building as well, part of its quickly growing Russian population.

Living among the new Russians, Mrs. Gumpertz was out of her element. In 1884, she inherited the fantastic sum of $600 from her husband's family in Germany and used the money to finance her move to Yorkville, a predominantly German neighborhood on Manhattan's Upper East Side. She remained in Yorkville, living with her daughters until her death in 1894. She was fifty-eight years old.

CHAPTER FOUR

The Rogarshevsky Family

In the first decade of the twentieth century, immigrant traffic between Europe and the United States reached its peak, with 1,285,349 immigrants arriving in 1907 alone, a number that stunned Americans at the time and has never been equaled since. A typical day in those high-volume years could see over three thousand immigrants pass through Ellis Island, most ferried to the mainland within two to three hours. Roughly ten percent, however, were detained as captive guests of the immigration authorities. Among them were eight members of the Rogarshevsky family, two adults and six children. The Rogarshevskys immigrated to the United States from Telsh, Lithuania, a town famous in the nineteenth century as a center of Jewish learning. Abraham and Fannie Rogarshevsky, their five children, along with an orphaned infant niece, sailed from Hamburg, landing at Ellis Island on July 19, 1901. Here, they were briefly detained. The reason given in the official documents was very simply "no money." The problem was most likely resolved by a relative who came to Ellis Island to vouch for the family, promising to support the Rogarshevskys until they found steady employment.

The Rogarshevskys were held for only a couple of days, but thousands of new arrivals found themselves stuck on Ellis Island for weeks and even months. The detainees fell into three basic groups. Women traveling alone were held on Ellis Island until a male relative came to fetch them, most often a husband or a brother. Another group contained the family members of immigrants held in the Ellis Island hospital. The final and most amorphous group was made up of immigrants who were "not clearly and beyond a doubt entitled to land." Deportees were also held on Ellis Island pending their return to whatever country they had come from. The vast majority of deportees were rejected as "paupers."

Detainees were housed in dormitories large enough to accommodate three thousand people. As the newspapers pointed out, that was more than the Waldorf-Astoria and Astor hotels combined. Unlike the Waldorf, however, the immigrant "hotel" on Ellis Island was a strictly no-frills operation. Guests slept on three-tiered bunks with wire mattresses, the bunks enclosed in pens that resembled oversized birdcages. Each morning, the pens were unlocked and disinfected to prevent the spread of typhus, cholera, and lice.

Along with shelter, Ellis Island provided new arrivals with nourishment: three meals a day served in a vast hall—the "world's largest restaurant," as one visitor described it. Diners sat at long bench-lined tables draped in sheets of white paper. In the interest of conserving space, the aisles between the tables were just wide enough for a grown man to squeeze through sideways. Even so, the immigrants ate in shifts, a thousand at a time, the first meal of the day served at half past five in the morning. Waiters in white jackets brought the immigrants their food. For many diners, it was the first time they had ever eaten food prepared and served by strangers.

Visitors to the mess hall were shocked by the immigrants' disregard for table etiquette: They dove into their food like birds of prey and tossed the scraps—the bones and potato peels—onto the floor. When the dining room was expanded in 1908, easy clean-up was factored into the new design. The entire space was covered in white tile and enamel paint, with every sharp angle or edge softened into a curve to prevent dirt from settling into the corners and crevices. The dining-room floor was

The immigrants' dining room at Ellis Island, date unknown.

sloped toward half a dozen drains, so the room could be easily hosed. "It is doubtful," one visitor concluded, "if the guests of any hotel in the country have their meals served under more satisfactory conditions of cleanliness, healthfulness, and good cheer."[1] As to the quality of the food, opinions were decidedly mixed.

Like the baggage-handlers and money-changers, the Ellis Island food purveyors were private contractors granted the privilege of doing business on government property, hence their generic title: "privilege holders." Of all the island's concessions, feeding the immigrants was the most lucrative, and local caterers competed for the job in public auctions. The results were announced in the local papers, like the final score in a sporting event. Along with running the dining room, the food concessionaire operated a lunch stand, where immigrants paid cash for bread, sausage, tins of sardines, fruit, and other portable items. In the dining room, the immigrant ate for free, the food paid for by the steamship companies that brought them to America. In 1902, that came to 35 cents a day for breakfast, lunch, and dinner, a small sum that added up quickly. During the high-volume years, feeding the immigrants detained on Ellis Island cost the steamship companies half a million dollars annually, but the money came out of the terrific profits they made on their steerage passengers, the golden goose of the shipping industry.

The immigrants' first lesson in American food ways, however, took place before they had even landed. Once their ship had docked, the immigrants were loaded onto barges that ferried them to Ellis Island. It was here that each passenger was handed a cup of cider and a small round pie, the quintessential fast food of turn-of-the-century America. The two foods that most impressed the new immigrants were bananas (many tried to gnaw through the skin) and sandwiches. As they waited their turn in the Ellis Island registry line, sometimes a thousand people long, waiters snaked through the crowd, distributing coffee and ham or corned-beef sandwiches. The immigrants munched appreciatively, marveling over the sweetness of American white bread.

The regimen in the Ellis Island dining room was meager and repetitive, a step up from prison fare. For breakfast, there was bread and bowls of coffee with milk and sugar. At lunch, the immigrants were given soup, boiled beef, and potatoes. For supper, more bread, this time with the addition of stewed prunes. Unscrupulous caterers and crooked officials conspired to winnow the big-ticket items (the meat and the dairy) from the immigrants' diet until all that was left was bread, coffee, and prunes. As a result, thousands of immigrants sustained themselves on an innovation of the Ellis Island kitchen: the prune sandwich.

In 1903, President Roosevelt launched an investigation into corruption on Ellis Island, which ended with a thorough overhaul of the reigning administration. One beneficiary of the regime change was the immigrant dining room. Menus tell the story best. The one below is from a later period, but captures the reformers' culinary mandate:

SUNDAY, JULY 1, 1917
BILL OF FARE
FOR THE
IMMIGRANT DINING ROOM

BREAKFAST
Rice with Milk and sugar
served in soup plates
Stewed Prunes
Bread and butter
Coffee (tea on request)
Milk and crackers for children

DINNER
Beef Broth with Barley
Roast Beef
Lima Beans–Potatoes
Bread and Butter
Milk and crackers for children

SUPPER
Hamburger Steak, Onion Sauce
Bread and butter
Tea (Coffee or Milk)
Milk and crackers for children[2]

The immigrants also dined on pork and beans, beef hash, corned beef with cabbage and potatoes, Yankee pot roast, and boiled mutton with brown gravy. These were the sturdy foods of the American working person served in accordance with the nutritional wisdom of the day. Cooked cereals, cheap but nourishing, were routine at breakfast, while the midday meal, the most substantial of the day, was built around protein and starch. Milk, the all-American wonder food, was available at every meal for immigrant children, and was freely dispensed between meals as well. Vegetables were more or less limited to peas, beans, and cabbage.

Given the very limited diets the newcomers were accustomed to, the great quantities of food that materialized each day in the Ellis Island dining room was cause for euphoria. The fact that it was free of charge was literally beyond belief. To reassure the immigrants, signs were posted in the dining hall in English, German, Italian, French, and Yiddish: "No charge for food here." Milk, bread and butter, coffee with sugar, all of it free and in endless supply. And the meat! A single day's ration on Ellis Island was more than many immigrants consumed in a month. The bounty of Ellis Island hinted at the edible riches that waited on the mainland. At the same time, the island also fed tens of thousands of waiting deportees, people who would never reach the mainland but were granted a fleeting taste of American abundance. Deportees spent their days locked up in holding pens, but the food they received on the island was wholesome and plentiful. According to one island official, the thick slabs of buttered bread and hot stews were so much better than any food the deportees had ever known that they wept at the thought of leaving Ellis Island, even if staying meant a lifetime of confinement.

Each year, on the last Thursday of November, detainees celebrated American abundance at a Thanksgiving banquet that featured roast turkey, cranberry sauce, and sweet potatoes. Whether or not they grasped the meaning behind the meal, the immigrants were clearly swept up in the festive spirit of the day. In place of flowers, the women bedecked themselves with sprigs of celery plucked from the tables, while the children feasted on candy and oranges. After the dinner was served, the men puffed on cigars, a habit acquired just for the occasion. The meal itself lasted for several hours, the waiters instructed to keep filling the plates until every diner was fully sated. When it was over, the immigrants were serenaded by a hundred members of a German singing society. Their final number was the *Star-Spangled Banner*, a song the audience had never heard before, sung in a language it couldn't comprehend. Nonetheless, the immigrants caught on quickly and rose to their feet, their heads bowed.

As to the food, it was also unfamiliar. The great majority of the guests had never seen a cranberry or an orange-fleshed potato, but the dish that perplexed them most was mince pie. A reporter from the *New York Sun* who visited Ellis Island in 1905 witnessed the immigrants' first tentative bite of this holiday classic:

> *Mince pie was a novelty as to form if not to contents to everyone who sat down to his first Thanksgiving dinner. Half a pie was served to each, but it was some minutes before the diners could make up their minds as to what they were getting and as to whether they would risk it.*

But then:

> *When once they buried their teeth in the spicy filling, it was easy to see that they would be willing converts to the great American practice of pie-eating.[3]*

The implications were clear. In that moment of conversion, their taste buds adjusting to the fruity richness, a future American was born.

Images of Ellis Island as a floating cornucopia contrasted sharply with the "island of tears" portrayed in the immigrant press, among the institution's most vocal critics. Foreign-language newspapers condemned Ellis Island for its overcrowding, its callous handling of new arrivals, and its overzealous implementation of immigrant law. When the complaints grew loud enough, government commissions were convened to investigate the charges. (Roosevelt's 1903 investigation was in response to a series of condemnatory articles that ran in the German-language newspaper, the New York *Staats-Zeitung*.) While some claims were exaggerated, many charges leveled by foreign-born reporters were essentially true. Immigrants were denied entrance to the United States on petty technicalities; they were treated with gruff indifference by island employees and wedged into bug-infested dormitories. The deeper truth, however, is that the brutal efficiency of the Ellis Island machine somehow coexisted with genuine attempts at humane handling of the alien masses.

Detention on Ellis Island was a dreary, physically demanding, and anxiety-ridden experience. During that first busy decade, the immigrants' dining room was among the island's only bright spots. (Another was the roof garden complete with boxes of flowering geraniums, awnings for shade, benches for resting, and a children's playground.) Over time, however, the men who ran Ellis Island looked to the immigrant depot as the first all-important point of contact between the United States government and its future citizens, developing a near-mystical belief in the power of that first encounter. Frederick Wallis, immigration commissioner from 1920 to 1921, summed up the new thinking this way: "You can make an immigrant an anarchist overnight at Ellis Island, but with the right kind of treatment you can also start him on the way to glorious citizenship. It is first impressions that matter most."[4] In his efforts to ensure the best possible impression, the commissioner introduced a series of reforms, imposing higher standards of cleanliness and courtesy. He established a baby nursery for young mothers, a playroom for children, and a recreation hall for adults. On weeknights, the immigrants attended lectures and motion-picture showings, while Sunday afternoons were set aside for live concerts. In the dining room, the new spirit of hospitality meant a more inclusive kitchen pantry, an attempt to satisfy the immigrants' diverse culinary needs. One of the most important additions to the Ellis Island regimen was pasta—or "macaroni," as it was listed on the menu. As the officials in charge of Ellis Island grew more attuned to the immigrants' native food customs, the job of feeding them grew more complex. But while each group traveled with its own set of culinary biases and food taboos, no group arrived with more stringent and elaborate dietary restrictions than the Jews.

According to government records, 1,028,588 Jews immigrated to the United States between 1900 and 1910. Of that number, the great majority came from the "Pale of Jewish settlement," a geographic designation created by Catherine the Great in 1791. Catherine established the Pale in an effort to corral and isolate the Jews living within the newly expanded Russian Empire. On a map, the territory corresponds to modern-day Ukraine, Poland, Latvia, Lithuania, Moldova, and Belarus. Russian Jews were well acquainted with anti-Semitism, but the Jewish mass migration that began in the 1880s was sparked by a wave of pogroms that heightened the Jews' perennial status as outsiders. They began in 1881 in what is now Ukraine, as rioting mobs destroyed millions of rubles' worth of Jewish property, killing dozens of Jews in the process. Small-scale pogroms continued for the next twenty years, erupting in full force in the city of Kishinev on Easter Day, 1903, when fifty Jews were killed during several days of uncontrolled violence. America promised Jews a safe harbor and political and religious freedom, along with unbounded economic opportunity.

Russian and Eastern European Jews lived primarily in small market towns known as shtetlach. Once a week, Gentile farmers from the surrounding countryside would converge on the shtetl to sell their goods and buy supplies from the Jewish shopkeepers, though shtetl Jews worked in many other occupations as well. The distinct folk culture that developed in the shtetlach found expression in language, music, and religion. Unlike their German brothers and sisters, shtetl Jews practiced an unambiguously traditional version of Judaism. Where men expressed their piety through study and prayer, women spoke through the language of food. The sacred responsibility of the shtetl homemaker was to keep a kosher home, celebrating the holidays with all the required ritual dishes. On the Sabbath, and other holy days, she distributed food to the poor. Landing on Ellis Island, these same Jews found a profusion of food, but, with a few exceptions, none of it was kosher.

Actually, the Jews' culinary problems started at sea. Though the steamship companies were legally obliged to feed their passengers, only a fraction served kosher meals. Some fulfilled the requirement with a single food: herring. Others went through the trouble of installing kosher kitchens but hired cooks who were kashruth-illiterate, unfamiliar with the full sweep of Jewish dietary law. Jewish travelers who knew what to expect traveled with their own survival rations. One very common food was thick slices of zwieback-like bread that had been dried in the oven to keep it from spoiling. Travelers also preserved their bread by dipping it in vinegar and sugar then baking it. For protein, they packed dried fish and salami. The chief problem with home-packed food was that it often ran out before the ship reached America, leaving the immigrant with nothing but water and perhaps some tea for the last leg of the journey. Between the rampant seasickness and the germ-infested quarters, no one in steerage—Jew or Gentile—fared particularly well. The Jews, however, faced the added challenge of finding kosher nourishment, an often impossible task, and many arrived at Ellis Island stooped with exhaustion, colorless, and malnourished. Unfortunately for them, the relief of standing on solid ground was quickly followed by another realization: there was still nothing to eat.

The one place freshly landed Jews could find nourishment was at the Ellis Island lunch stand, which carried tinned sardines and kosher sausages. But the stand was only accessible to Jews who had already passed inspection. For Jews detained on the island, the food situation was grim. There was

The food counter at Ellis Island, 1901.

nothing kosher about the immigrants' dining room, which left devout
Jews with a choice: they could either go hungry and possibly starve, or
break the food commandments and eat. (The Rogarshevsky family faced
this precise dilemma in 1901, though only briefly.) The one time of year
Jews were assured of a good kosher meal was at Passover, the springtime
feast commemorating the Hebrew exodus. Under the headline "Passover
at Ellis Island," in 1904, the *New York Times* ran this short but evocative
story on the immigrants' seder:

> The Feast of the Passover was celebrated in due form last night at
> Ellis Island by 300 Jewish immigrants, detained there awaiting in-
> spection. Commissioner Williams gave them permission to celebrate
> the rites of their church and the great dining hall was turned over to
> them, and there, dinner was served in keeping with the occasion.
>
> The tables were covered with snowy linen and new dishes right
> from the storeroom. In the kitchen, the utensils were all new, and the
> dinner was cooked under the supervision of the immigrants them-
> selves. The dinner was rather more sumptuous than is usually served
> to incomers—chicken soup, roast goose and apple sauce, mashed pota-
> toes, ground horseradish, matzoth, black tea, and oranges.[5]

A gastronomic retelling of the Jews' escape from slavery, the Passover
meal held special significance for the immigrants. The parallels were perfectly clear: Russia was their Egypt, the czars were their pharaohs, while
America was their modern-day Canaan. But Passover came just once a
year.

Relief for the kosher food drought on Ellis Island arrived in 1911,
when the Hebrew Immigrant Aid Society finally convinced the authorities to give the depot its own kosher kitchen. HIAS, as it was known,
was just one of the many immigrant aid societies with offices on Ellis
Island, each one serving the needs of a particular ethnic or national
group. Founded on the Lower East Side in 1902, HIAS was formed
with a single mandate: to provide any Jew unlucky enough to die on
Ellis Island with a proper Jewish burial. From that narrow focus, the

mission quickly broadened to helping new arrivals gain a firm foothold in their adopted country. HIAS representatives wearing blue caps with the HIAS acronym embroidered in Yiddish met the incoming ferries and distributed pamphlets (also in Yiddish) on the inspection process. They helped steer immigrants through the island's bureaucratic maze and advocated for immigrants condemned to return to "the country from whence they came."

From their vantage point on Ellis Island, it was plain to the HIAS workers that the kosher-food shortage diminished the immigrant's chance of passing inspection. The Ellis Island doctors sorted all new arrivals into classes, admitting most but barring anyone with tuberculosis, epilepsy, or any other "loathsome and dangerous disease." In their decrepit post-voyage state, a high percentage of Jews fell into the catchall category "LOPD," bureaucratic shorthand for "lack of physical development." It was a vague diagnosis, and not especially loathsome, but serious enough to block the immigrant from entering the country. The reason was purely economic. According to the Ellis Island calculus, physical weakness diminished the individual's earning power, a most serious consideration. Along with "pauper," the single largest class of unwanted foreigners, the languishing Jews were officially rejected with another catchall label, LPC, or "likely to become a public charge," when all they really needed was a few square meals and a good night's rest.

In 1911, a New Yorker named Harry Fishel took this argument to Washington and presented it to President Taft. An immigrant himself, Fishel was a Donald Trump–like figure who made his fortune in the New York real estate market, purchasing and developing large tracts of land. Many of his holdings were on the Lower East Side, including one entire block of tenements on Jefferson Street. Fishel was also an Orthodox Jew who had channeled his wealth into yeshivas, hospitals, and assorted charities, including HIAS, where he served as treasurer for over half a century.

Harry Fishel's crusade to feed the immigrants was doubly motivated. An act of compassion on behalf of the helpless foreigner, it was also an act of self-preservation. The way Fishel saw things, the kosher-food predicament on Ellis Island served as a roadblock to the kind of Jews

America needed most, the rabbis and scholars who were so essential to the future survival of Orthodox Judaism in secular America. Here was a cause the mogul was ready to fight for. Face-to-face with the president, Fishel pleaded his case with the urgency of a condemned man. He returned to New York the following day with a firm pledge that the United States government would do what it could to fill the kosher gap.

The food served in the kosher dining room was instantly recognizable to the immigrant palate. There were kippered herring, noodle and potato kugels, barley soup, and dill pickles. American specialties also made regular appearances. The following menus from 1914 are the earliest on record:

MONDAY

BREAKFAST:

Boiled eggs (2)
Bread and butter
Coffee

DINNER:

Potato soup
Hungarian goulash
Vegetables
Bread

SUPPER:

Pickled herring
Fresh fruit
Bread and butter
Tea

TUESDAY

BREAKFAST:

Fresh fruit
American cheese

Bread and butter
Coffee

DINNER:

Vegetable soup
Pot roast
Potatoes
Bread

SUPPER:

Bologna
Dill pickles or sauerkraut
Stewed fruit
Bread and tea

WEDNESDAY

BREAKFAST:

Fruit
American sardines
Bread and butter
Coffee

DINNER:

Barley soup
Roast meat
Vegetables
Bread

SUPPER:

Beans (baked by Mrs. Paley)
Cakes
Bread and tea[6]

The size of the crowd in the dining room rose and fell depending on that day's shipping schedule. The room might be empty for breakfast, but if a ship arrived that afternoon filled with Russians or Hungarians or Poles, the kosher kitchen went into high alert, capable of feeding six hundred mouths at a single sitting. One interesting footnote to the Ellis Island kitchen's history is the leading role played by women. During World War I, a woman known to us only as Mrs. Paley (first initial "S") was in charge, but in later years the job of head cook fell to another woman, whom we know much more about.

When Sadie Schultz came to work on Ellis Island in 1929, she was forty-six years old with a long culinary résumé. Born in Canada in 1882, Sadie Citron Schultz was the daughter of Polish immigrants who had returned to Europe when Sadie was still a young girl and then re-emigrated to the United States, settling on the Lower East Side. According to family legend, Sadie's mother earned the family's passage working as a cook for one of the steamship companies, which helps explain the zigs and zags in their route to America. The young Sadie Schultz entered the New York food economy at twelve years old with a waitressing job in an East Side restaurant. She worked steadily from that point on, with only two interruptions. The first was in 1906, the second in 1910, the years her children were born. As soon as the babies were old enough, she put them into the free nursery at the Educational Alliance on East Broadway and returned to the restaurant. At some point in her career, she traded in waitressing for a job behind the stove, and this is where she remained for the rest of her professional life.

Under Mrs. Schultz's command, the kosher dining room functioned like a home kitchen writ large. Despite the institutional scale of her work, Mrs. Schultz worked without the benefit of written recipes. Recipes would, in fact, have been useless to her, as she could neither read nor write. The dishes she prepared for her immigrant clients were the same ones she made for her own family, the standard offerings of the Jewish home cook. Even more, Mrs. Schultz learned the regional food preferences of each national group and tailored her cooking to suit their tastes, the same way mothers adapt their cooking for a finicky child. So, for example, if she learned that Ellis Island was expecting a boatload of Hungarians, she prepared her stuffed cabbage with raisins and sugar, to satisfy the Hungarian sweet tooth. For Lithuanians, she omitted the sugar and added vinegar.

The following stuffed-cabbage recipe comes to us from Frieda Schwartz, born on the Lower East Side in 1918. Her special touch is the addition of grated apple to the filling.

STUFFED CABBAGE

1 lb beef

1 egg

3 cups canned tomatoes

½ tsp pepper

Beef bones

1 peeled and grated apple

3 tbsp rice

3 tbsp cold water

4 tbsp grated onion

3 tsp salt

1 cabbage

Pour boiling water over cabbage. Let stand 15 minutes. Separate the leaves. Remove the thick stem from the outside of each leaf. Prepare the sauce in a heavy saucepan by combining tomatoes, salt, pepper, and bones. Cook 30 minutes, covered. Mix beef, rice, onion, egg, apple, and water. Place a heaping tablespoon of the mixture in a cabbage leaf. Roll leaf around mixture and add to sauce. Season with lemon juice and brown sugar. Cook 2 hours.[7]

Fannie Rogarshevsky and her son Philip, circa 1917.

Fannie Rogarshevsky gave birth to two more children in America, bringing the total to six: two girls and four boys. For a time, the Rogarshevskys
lived at 132 Orchard Street, moving down the block to number 97
sometime around 1908. The building was also home to Fannie's parents,
Annie and Joseph Beyer, who had adopted their orphaned granddaughter.
According to the 1910 census, the sixty-four-year-old Mr. Beyer earned
his living as a street peddler. During the 1920s, another set of relatives,
the Bergmans, lived directly across the courtyard from the Rogarshevskys,
the two buildings connected by a clothesline. Mrs. Rogarshevsky occasionally used the clothesline as a delivery system, sending the Bergmans
pots of cholent, a Sabbath stew.

At the time of the 1910 census, all six of the Rogarshevsky children
were still living at home. Ida, age eighteen, was employed as a "joiner"
in a garment factory; Bessie, age sixteen, was a sewing-machine operator;
and Morris, age fifteen, worked as a shipping clerk. Sam and Henry, ages
twelve and seven, were in school, and three-year-old Philip was at home

with his mother. The four Rogarshevsky boys slept in the parlor room on a jury-rigged bed, their heads resting on the sofa, their feet supported by four kitchen chairs. The two girls shared a folding cot, while their parents slept in the "dark room," on the other side of the kitchen. The Rogarshevsky boys spent their free time haunting the front stoop and getting into street fights. (One of them, Sam, trained in a local boxing gym with the hopes of going professional, a career many East Side boys dreamed about and some achieved.)

On the 1901 ship's manifest, Abraham Rogarshevsky described himself as a "merchant." In New York, however, he worked as a presser in a garment factory, a job held exclusively by men. Mr. Rogarshevsky was paid "by the piece," his salary dependent on the number of garments completed per week, a common point of contention between factory workers and managers. By 1917, Mr. Rogarshevsky had been diagnosed with tuberculosis, which was then known as the "tailor's disease," and died the following year. After her husband's death, Fannie supported her family by taking in boarders, a common recourse of East Side widows. At the same time, she became the building's janitor, a job that offered no wage but allowed her to live there rent-free. Between her two jobs, the income of her older children, and the kindness of her neighbors, the single mother of six cobbled together a decent existence. Among her allies were the East Side peddlers, so indispensable to neighborhood homemakers, who provided a wide array of edible goods at the lowest prices in the city.

For Eastern European Jews, the city's pushcart markets were a reminder of home. The shtetlach that the immigrants had left behind had one key feature in common: an outdoor food market where, once a week, Jewish homemakers shopped for their supplies. Some of what they needed could be found in stores, but they relied on the market for their produce, their poultry, fish, milk, cheese, and butter, as well as household goods like candles, pots, and pans. When they arrived in New York, they found themselves perfectly at home in the tumult of the East Side push-cart markets that had been created by their immigrant predecessors.

To uptown visitors, the East Side pushcart markets were garbage-strewn streets aswirl with foreigners, women in tattered wigs, baskets over one arm, haggling at top volume over third-rate merchandise. In other words, retail mayhem. The more Bohemian of the uptown visitors swooned over the romance of the pushcarts. They came to the markets as sightseers (never as customers) to drink in the Old World atmosphere and observe the local customs. Whenever the city threatened to close down the markets, which it did at regular intervals, they mourned the impending loss, mostly on aesthetic grounds. The pushcart markets on Hester, Orchard, and Essex streets were among the most picturesque spots in New York, and the city would be a much grayer place without them. The point that seemed to elude them was the usefulness of the markets to the people they served. For the tenement housewife, the pushcarts were America's antidote to hunger. They provided her with a wide assortment of familiar foods at the lowest possible prices, and allowed her to buy them in the quantities she desired. Where else in New York could she buy half a parsnip or a handful of barley, not a single ounce more than she needed? The minuscule purchases possible at the pushcarts surprised uptown New Yorkers, who wondered why anyone would buy a single egg, but to the tenement housewife, it was eminently practical. She had no pantry to store her provisions and no ice box to keep foods from spoiling. More compelling still, small purchases were the only kind she could afford.

Tenement housewives like Fannie Rogarshevsky shuttled between the pushcart and the kitchen at least twice a day. In the mornings, before the children were awake, they bought their breakfast supplies, some hard rolls and maybe a cup of pot cheese. In the afternoon, they returned to the market for their dinner ingredients. To the uptown city-dweller, the idea of shopping meal-by-meal was hopelessly inefficient. The tenement housewife saw things differently, treating the pushcarts as an extension of her own kitchen. For Mrs. Rogarshevsky, who lived directly above the Orchard Street market, this was almost literally the case, and the same was true for thousands of other East Side women.

The pushcart market was a boon to East Siders on both sides of the equation, shopper and peddler alike. A line of work familiar to the

Eastern European Jew, peddling was the fallback occupation of new immigrants. It required little capital, no special work skills, and scant knowledge of English. All immigrants needed were a basket and a few dollars to invest. Many started with dry goods—suspenders, collar buttons, sewing pins, and the like—which they peddled door to door. The pushcart, a larger retail venue, demanded more capital and a deeper knowledge of the workings of the city. Pushcart peddlers rented their carts for 10 cents a day from one of the many East Side garages or pushcart stables. They began work each morning around four a.m., wheeling the carts to a wholesale produce market on Catherine Slip along the East River, which catered specifically to the pushcart trade. By five a.m., carts loaded, they were on the street and ready for business. At some point in the afternoon, the peddlers' wives took over the cart so the men could rest up for the next day's early start. (Actually, a fair percentage of peddlers were women, and only some were partners with their husbands.) The chief attraction of peddling for the Eastern European Jews was the independent nature of the work. The sweatshop worker had precise hours to keep, quotas to meet, and supervisors to appease. The peddler, by contrast, was his own boss. As one East Sider put it, "the peddler was a man who had seen the sweatshops and thought they were for someone else." There was dignity in peddling, but, even more to the point, the peddler was free to set his own hours and keep the Sabbath. To observant Jews, this was a crucial advantage over the sweatshops, which followed the Gentile business week and stayed open Monday through Saturday.

Beginning in the 1890s, the pushcart market became a regular destination for New York journalists, who were lured by its literary possibilities. They came in search of good copy and found it in characters like the Polish fishmonger with her barrels of two-penny herrings, or the horseradish peddler, bent over his mechanical grinder, literally reduced to tears by the rising fumes. A more quantitative rendering of the pushcart market was provided by New York mayor George B. McClellan, who presided over City Hall from 1904 to 1909. Pressured by public concern over the quickly growing number of pushcarts, Mayor McClellan appointed a commission to investigate what some New Yorkers referred to as "the pushcart evil." Their complaints were many. The pushcarts, they said, were a threat to public health. They generated garbage and interfered with proper street-cleaning. They sold contaminated food—moldy bread, worm-ridden cheese, rotten produce—to New York's most vulnerable citizens. Even more pressing, the pushcarts interfered with the free flow of traffic in a rapidly expanding metropolis, a matter of great concern to city officials.

To establish a common body of facts, Mayor McClellan ordered a systematic "pushcart census," and on May 11, 1905, a small army of police officers fanned out over the Lower East Side, each one armed with a stack of questionnaires. To some measure, the census confirmed what everybody already knew. The one neighborhood with more push-carts than any other was unequivocally the Jewish ghetto. Of the four thousand pushcarts counted in Manhattan, two thousand five hundred were on the Lower East Side, with the highest concentration on Hester, Orchard, and Essex streets. The census also brought surprises. The push-cart peddlers earned a better living than anyone suspected, and stayed in their jobs longer than anticipated. Peddling was not just a stepping-stone job, as most people believed, but a destination. Another surprise was the high quality of the goods. Contrary to expectation, more than 90 per-cent of the fruits, vegetables, eggs, butter, cheese, and bread sold from the pushcarts was declared fresh and wholesome, of better quality than the same items found in a store. The public world of the market offers a rare glimpse into the private realm of the kitchen. Thanks to the mayor's census, we know precisely what foods were available to the tenement housewife and which she relied on most.

Health workers who studied the immigrants' eating habits in the early part of the twentieth century bemoaned the shortage of vegetables on the Jewish dinner table. The ghetto market, however, abounded with vegetable peddlers. Of course, there were potatoes, but there were also beets, cabbage, carrots, eggplant, parsnip, parsley, rhubarb, onions, pep-pers, peas, beans, cucumbers, radishes, and a food listed as "salad greens." One reason the health workers may have overlooked Jewish vegetable consumption is that so much of it came in the form of soup.

There's an old Yiddish proverb that goes: "Poor people cook with a lot of water." The truth of the proverb was borne out on a daily basis in the immigrant soup pot. In the winter months, Jewish cooks like Mrs. Rogarshevsky prepared tangy, magenta-colored borschts; cabbage soup; chicken soup with carrots, celery, and parsnip; potato soup enriched with milk; and, most economical of all, bean soup, a dish found throughout the tenement district regardless of the cook's religion or country of origin. Lima beans, fava beans, white beans, lentils, chickpeas, and dried peas both yellow and green were cheap, nutritious, and easy to cook. Jewish cooks liked to combine their beans with onion, carrot, celery, and barley, producing soups that were deeply flavored and slightly chewy. They called the soup *krupnik*, a dish traditionally served to impoverished yeshiva students. In its simplest form, *krupnik* was indeed a spartan dish, nothing more than lima beans, a handful of barley, and maybe a chunk of potato. Adding a marrow bone was one way to make it more substantial. For meatless *krupniks*, the cook might add a splash of milk or maybe some dried mushrooms, an ingredient that mimicked the savoriness of meat.

In the mid-1930s, the *Daily Forward*, the East Side's leading Yiddish newspaper, began a regular cooking feature edited by Regina Frishwasser. The recipes that appeared in the column were sent in by readers—home cooks with limited time and limited budgets as well. In the 1940s, Frishwasser collected the recipes into *Jewish American Cook Book*. The purpose of the book, she writes in her preface, "is not to bring glamour to a menu, but rather to bring our foods in the easiest way possible to those who want them." Here is her recipe for a *krupnik* that used dried mushrooms, barley, lima beans, and yellow split peas.

KRUPNIK

Bring 2 quarts water to a boil, and add 1 cup yellow split peas, ½ cup minute barley, ½ cup lima beans, and 1 teaspoon salt. Simmer 1 hour and add 1 ounce broken dried mushrooms, 1 minced onion, 1 diced carrot, and 1 diced parsley [root]. Cook until the vegetables are tender. Fry I minced onion in 2 tablespoons butter until golden brown, then add to the soup.[8]

———

Come summer, Jewish cooks turned to chilled soups, like meatless borscht served with sour cream and boiled egg, just one of the many mouth-puckering foods consumed by the immigrants, a taste preference they had acquired on the other side of the ocean. Back in Europe, the traditional souring agent in borscht was home-fermented beet juice otherwise known as *rossel*. Once in America, cooks turned to a store-bought product called sour salt (tartaric acid) to give their borscht the required zing. Like lemonade, it was the sourness of borscht that made it so refreshing. Schav was another cold and sour soup that the Jews consumed as a summer tonic. Murky green in color, it was made from boiled and chopped sorrel leaves, a plant loaded with vitamin C. The appearance of sorrel on the East Side pushcarts signaled that spring had come to the ghetto. Tenement housewives prepared their first batch of schav sometime in mid-May, and served it "the old Ghetto way," with sour cream, bits of chopped egg, cucumber, and scallion, so it was part soup and part salad. Schav was also popular in the East Side cafés, where customers sipped it from a glass like iced tea.

In warm weather, as pushcarts filled with summer vegetables, the Jews became avid salad-eaters, though not the leafy green kind favored by the Italians that we are most familiar with today. Instead, they chopped cucumber, radish, scallion, and pepper into bite-size chunks and sprinkled them with a little salt and pepper. In a more luxurious version, the raw vegetables were crowned with a scoop of cottage cheese or sour cream, a dish once referred to as "farmer's chop suey." This classic Jewish creation was reportedly the food that Harry Houdini (a Hungarian-born Jew) requested on his deathbed.

When the first pushcart survey was taken in 1905, fruit peddlers held sway over the market, occupying more curb space than vendors of any other food. On the far side of the ocean, Jewish fruit consumption was

Family members often took turns at the pushcart. Children peddled in the afternoon when school let out.

more or less limited to whatever grew locally, including apples, peaches, cherries, berries, and, above all, plums, which grew on the outskirts of the shtetls and which Jewish cooks made into a thick, dark preserve called *pavel*, a kind of plum butter. Plums were also dried along with apples and used in cooking. Jewish cooks added prunes to festive dishes like tzimmes (sweet glazed carrots) and cholent, or used it as a filling for hamantaschen, the triangle-shaped Purim cookie. When crushed and left to ferment, plums were the foundation for slivovitz, a kind of Eastern European firewater. At the pushcart market, immigrant Jews discovered an Eden of melons, citrus, stone fruits, and tropical wonderments like pineapple, banana, and even coconut, which the vendor sold, pre-cracked, the white oily shards floating in jars of cloudy water. In fact, many kinds of fruits—melons, pineapple, even oranges—were sold presliced and hawked as street food, a practice that city officials frowned on. (According to the New York sanitary police, the consumption of bad fruit purchased from street peddlers was a leading cause of death among East Side children.) Where other vendors packed up by dinnertime, the fruit vendors remained on the street long after the sun went down, their carts illuminated by flaming torches. Fathers coming home from work would stop by the fruit peddler for penny apples to give to the kids. On summer nights, when tenement-dwellers poured into the streets for a breath of fresh air, strolling East Siders paused at the fruit carts for a cool slice of watermelon. Fruit was the great affordable luxury of the tenement Jews.

In the early 1920s, a Boston dietician named Bertha Wood conducted a multiethnic study of immigrant eating habits, eventually published as a book, *Foods of the Foreign-Born in Relation to Health*. As the title suggests, the book was written for health-care professionals—visiting nurses, settlement workers, and dispensary doctors—who served the immigrant community. Though well versed in current medical practice, they knew very little about the immigrants' foodways, a tremendous handicap in treating the immigrant patient. For each group in her study, Wood identified the leading food deficiencies and most harmful tendencies. She was also ready, however, to point out where the immigrant cook was superior to her native-born counterpart.

At less than a hundred pages, *Foods of the Foreign-Born* is a curious little book. Ms. Wood approaches her immigrant subjects with a degree of culinary open-mindedness unusual for the 1920s, a particularly anxious period in American political history. At the same time, she is firmly moored in the food wisdom of her day, with a deep faith in the value of bland, unadorned cooking like creamed soups and boiled vegetables. Her 1920s perspective helps explain Wood's two most persistent concerns with the immigrant kitchen: too much seasoning and too little milk. Ms. Wood declared the Jews guilty of both preparing highly seasoned foods (one reason the Jews were so nervous) and depriving their children of sufficient milk, "nature's most perfect food."

Red Cross workers distributing milk, "nature's most perfect food," to newly landed immigrants.

Among the foods that Ms. Wood objected to most was a much-loved Jewish staple: the pickle. "Perhaps no other people," Wood observed, "have so many 'sours' as the Jews. In the Jewish sections of our large cities," she continued,

> There are storekeepers whose only goods are pickles. They have cabbages pickled whole, shredded, or chopped and rolled in leaves; peppers pickled; also string beans; cucumbers, sour, half sour, and salted; beets; and many kinds of meat and fish. This excessive use of pickled foods destroys the taste for milder flavors, causes irritation, and renders assimilation more difficult.[9]

More alarming still was the pickle habit among Jewish school kids, who spent their lunch money on pickles and nothing else, their appetites ruined for more appropriate foods like milk and crackers. The taste of the standard Jewish pickle was so aggressive—briny, garlicky, sour—and so foreign to the native palate that Americans like Ms. Wood wondered how anyone, children especially, could eat them by choice. Instead, they saw pickle-eating as a kind of compulsion. The undernourished child was drawn to pickles the same way an adult was drawn to alcohol. More than a food, the pickle was a kind of drug for tenement children, who were still too young for whiskey.

At the pushcart market, the pickle stand was a rendezvous for shoppers. Here, standing among the barrels, hungry East Siders could buy a single pickle and eat it on the spot, then continue with their errands. Pickles were also sold in bulk, dished from the barrel with a sieve and packed into jars supplied by the shopper. Uptown visitors to the market were shocked by the size of Jewish pickles, some "large enough to kill a baby." These overgrown sours were cut into thick rounds that sold for a penny a piece and placed between bread to make a pickle sandwich, a typical East Side lunch.

The following recipe is adapted from Jennie Grossinger's *The Art of Jewish Cooking*:

DILL PICKLES

30 Kirby cucumbers of roughly the same size
½ cup kosher salt
2 quarts water
2 tablespoons white vinegar
4 cloves garlic
1 dried red pepper
¼ teaspoon mustard seed
2 coin-sized slices fresh horseradish
1 teaspoon mixed pickling spice
20 sprigs of dill

Wash and dry cucumbers and arrange them in a large jar
or two smaller jars, alternating a layer of cucumbers with
a layer of dill. Combine salt and water and bring to boil.
Turn off heat. Add vinegar and spices and pour liquid over
cucumbers. They should be immersed. If necessary, add
more saltwater. Cover and keep in a cool place for 1 week.
If you like green pickles, Mrs. Grossinger recommends you
try one after 5 days.[10]

Though pushcarts formed the backbone of the immigrant food economy, East Siders also patronized neighborhood shops: butchers, groceries, delicatessens, and dairy stores. This last group, a type of business that
no longer exists, included Breakstone & Levine, sellers of milk, butter,
and cheese, formerly located on Cherry Street, and forerunner to the
modern-day Breakstone brand. But inside the tenements, hidden from
the casual observer, immigrants trafficked in a shadow food economy in
which neighbors took responsibility for feeding each other. Transactions
within the tenement were most often cashless. Neighbors exchanged gifts
of food as part of an improvised bartering system in which the poor
gave to the truly destitute, or, in many cases, to families struck by tragedy: a death, sickness, a lost job. In return for her edible gifts, the tenement homemaker received the same consideration whenever her luck was
down—and no one in the tenements was immune from a run of bad
luck. Mrs. Rogarshevsky, a widow with six kids, was certainly eligible,
and edible charity must have streamed into the apartment during and
after her husband's long illness, when she adjusted to her new role as
breadwinner.

The continuous give-and-take that carried food from one apartment
to another was a strategy for survival among tenement-dwellers sustained
by the tenement itself. In buildings where apartment doors were hardly
ever locked or even closed, where stairways were used as vertical play-
grounds, rooftops functioned as communal bedrooms, and front stoops

Fannie Rogarshevsky (back row, second from right) and three of her children posing in front of 97 Orchard, circa 1920. (The two children in the front row are unidentified.)

were open-air living rooms, the business of daily life was an essentially shared experience. Tenement walls, thin to begin with, were riddled with windows, windows between rooms, between apartments, and windows that opened onto the hallway. As a result, sounds easily leaked out of one living space and into another. Or, if they were loud enough, ricocheted through the central stairwell. During summer, when East Siders hungered for fresh air, and windows to the outside world were open wide, voices were broadcast through the building via the airshaft. In the brownstones and apartment houses above 14th Street, New Yorkers lived more discreetly, sealed off from the larger world in their own domestic sanctuary. In the tenements, the people who lived above and below you were often blood relatives, but even if they weren't, you were fully briefed on their domestic status down to the most intimate details, and vice versa.

The communal nature of tenement-living was unavoidable and frequently unbearable. (Tenement-dwellers craved two things that many took

for granted: privacy and quiet.) At the same time, it encouraged neighbors
to look after each other in ways unheard of in other forms of urban housing. Visitors to the tenements, settlement workers, sociologists, and social
reformers, were struck by the generosity of the poorest New Yorkers,
recounting their many acts of kindness in memoirs and studies. In fact,
their writing became so cluttered with examples of tenement compassion
that Lillian Wald, founder of the Henry Street Settlement, was moved to
write, "it has become almost trite to speak of the kindness of the poor to
the poor," by way of introducing her own extended list.

The acts of kindness took many and varied forms. Between 1902 and
1904, a sociology student named Elsa Herzfeld carried out a study of
family life in the tenements, which outlined some of them:

> The readiness to share seems to me to be one of the chief traits in the
> relation of neighbor to neighbor. The aid given is of a simple kind. It
> satisfies an immediate need. Above all it is spontaneous. . . . When
> a mother has to go out for the day, she "leaves" the children with her
> neighbor or asks her to go in "and have a look at them." The neighbor
> comes in when the children are sick, she offers her blankets, makes some
> soup, suggests her own physician, or brings cakes and goodies to the
> sick child. She visits a neighbor patient in the hospital and brings her
> ice cream and candy or flowers. If a mother dies suddenly, a neighbor
> takes the children to her own room. If a child is neglected, she takes her
> "for months and asks no board." The young girl on the same floor is
> given a place in the home "to keep her from fallin' into low company."
> If your husband gets "drunk," a neighbor opens her door to you. If you
> get separated or dispossessed, "she has always room for one more."[11]

As a rule, East Siders avoided taking handouts from the established
charities because of the stigma it carried. Even the trip to the charity
office, oftentimes located in alien neighborhoods, was a much-dreaded
exercise in humiliation. In a coming-of-age memoir set on the Lower East
Side, Bella Spewack, who later went on to a successful career as a Broadway playwright, describes her mother, pregnant at the time and with two

kids already at home, abandoned by her husband, trying to convince the charity officer that she was worthy of $14 in rent money. "To get help from that place . . . you must cry and tear your hair and eat the dirt on the floor," Spewack writes. But seeking help from a neighbor was another story entirely. In the tenements, there was no need to explain or plead your case. Passengers on the same proverbial boat, the people around you grasped your situation with perfect clarity and gave what they could, with no probing questions or edifying lectures.

Sharing food with neighbors was standard practice among immigrants of every nationality, and in some cases, between nationalities. So, for example, an Italian housewife fed minestrone to the Irish kids who lived on the second floor, while Russians brought honey cake to the old Slovak lady across the airshaft. Widespread though it was, food-sharing loomed especially large among Jewish immigrants, who arrived in the tenements with their own long history of culinary charity. Fridays just before sundown, in the towns and cities they had come from, a woman who could afford the extra expense prepared a little more food than her own family needed and distributed it to less well-off neighbors. The sight of women carrying loaves of challah through the streets, or covered pots of stewed fish was a regular Friday-night occurrence. A second option was to invite a poor stranger to join the family Sabbath table, an old widower or beggar, or maybe a peddler who was miles from his own home. Public feasts were held for the poor in honor of weddings and brisses, the circumcision ceremony held on the infant's eighth day of life. Food-sharing, in short, was a built-in feature of the Jewish kitchen.

Fannie Cohen was an immigrant homemaker from Poland, who arrived in New York in 1912, a married woman with two young kids. Her husband was already here, having immigrated a full eight years earlier, and was working on the East Side as a carpenter. The family lived at 154 Ridge Street on the Lower East Side, in a tenement much like 97 Orchard, where Mrs. Cohen gave birth five more times, though one of the children died at fourteen months from contaminated milk. Mrs. Cohen had received a classic Jewish culinary education. Friday nights, she made gefilte fish (her standard formula combined whitefish, carp, and onion), which she chopped in a large wooden bowl and simmered along with the fish bones, wrapping them first in cheesecloth to prevent the kids from choking on one. She also made roasted carp, the whole fish rubbed with peanut oil, chopped garlic, and paprika, then baked until the skin was varnished-looking and slightly crisp. On Shavuot, the holiday in late spring that celebrates God's gift of the Ten Commandments, she made yellow pike, the fish sliced crosswise into meaty steaks then simmered with lemon, bay leaves, pepper, and a pinch of sugar. The main culinary attraction on the Purim table was goose and, for dessert, hamantaschen. On Passover she made brisket, fruit compote, and *chremsel*, dainty pancakes made from matzoh meal that were eaten with jam or dipped in sugar.

Whatever the holiday, it was Mrs. Cohen's habit to prepare more food than her own family could ever consume. Friday mornings at three a.m., she mixed up a batch of dough for the Sabbath challah, using twenty pounds of flour, forty eggs, and five cups of oil. The only vessel large enough to hold it was a freestanding baby's bathtub. Once mixed and kneaded, the dough was left to rise in its metal tub, covered by a wool blanket, until mid-morning, when it was sectioned off into loaves and left to rise again. By afternoon, Mrs. Cohen had twenty braided loaves cooling by the window. Some she gave to the neighbors, and some to the local rabbi, who always received the two largest loaves, each one the size of a placemat. Two loaves she kept for the family, but the rest she packed up and delivered to the newly landed immigrants at Battery Park, ferried there directly from Ellis Island. Knowing that some of them were stranded for the night—a tragedy on the Sabbath, when every Jew should be celebrating—she also came with soup, conveyed across town in metal canisters with screw-on tops.

In America, the newly arrived immigrant became the main recipient of the Jew's edible charity, while the tenement became the new shtetl. On Shavuot, Mrs. Cohen made hundreds of blintzes—some blueberry, some cheese, some potato—and sent them through the building, delivered by one of her kids. For Passover, she sent around tins of flourless sponge cake. But sometimes, there was no holiday at all and Mrs. Cohen still fed the building, handing a plate of stuffed cabbage or a square of kugel to one of the children with the instruction, "Bring up Mrs. Drimmer some food" or "Take this to Mrs. Sipelski," depending on which of the neighbors was sick or jobless or otherwise in need. These food deliveries always involved a round trip, since the child was later sent back to retrieve the now-empty dish. And sometimes the charity extended beyond the tenement to the larger neighborhood, like on Passover, when Mrs. Cohen invited strays, down-and-out characters whom her husband had rounded up on the way back from the synagogue, to eat from her own Seder table.

Here is Mrs. Cohen's challah recipe, scaled down to yield two good-size loaves:

CHALLAH

2 ½ lbs or 7 ½ cups bread flour

2 ounces fresh yeast or 4 teaspoons instant yeast

1 ½ cups warm water

½ cup peanut oil or other vegetable oil

4 eggs, room temperature

½ cup sugar

3 tablespoons salt

Dissolve yeast in warm water and let stand until mixture looks foamy, 5 to 10 minutes. Combine with remaining ingredients, stirring to form a dough. Knead dough for 10 minutes, then place in a lightly greased bowl. Cover with a damp cloth and let rise until doubled in volume, 1 to 2 hours. Punch down dough, knead ten times and divide in two. Separate each half into thirds. Roll each section into a rope about 18 inches long. Braid ropes, pinching the ends and turning them under. Place on a lightly greased baking tray and cover with a damp cloth. Let rise until doubled in size. Preheat oven to 375°F. Before baking, brush challah with one egg yolk mixed with 1 teaspoon water. Bake for 45 minutes to 1 hour, or until brown.[12]

Though food items of one kind or another dominated the pushcart trade, the market also provided East Siders with a sweeping array of nonedible goods. Pots, pans, dishes, scissors, soap, clothing, hats, and eyeglasses are just a minute sampling. In short, any useful item from mattresses to sewing thimbles was available at the pushcart market. But East Side vendors also trafficked in more fanciful goods, including the decorative objects known in Yiddish as tchotchkes, figurines, wax fruit, and mass-produced wall prints. The subject matter of pushcart artwork was often inspirational, the immigrant drawn to portraits of heroic figures from the worlds of literature and politics. There were postcard-size prints of William Shakespeare, Henrik Ibsen, Leo Tolstoy, Sholem Aleichem (the Yiddish Mark Twain), and President Lincoln, the great hero of the ghetto. Another revered figure was Christopher Columbus and, in later years, Franklin Roosevelt. Fannie Rogarshevsky kept a plaster bust of Columbus on the parlor mantel.

The market also supplied East Siders with intellectual stimulation in the form of books. Browsing the pushcarts for reading material was, to put it mildly, a hit-or-miss venture. The carts carried a grab bag of mostly secondhand volumes, many of them reference books. On the same cart, the shopper might come across a history of railway statutes, a yearbook from the Department of Agriculture, a collection of Hebrew prayer books, and an assortment of dictionaries. More discerning readers skipped the carts and shopped from the profusion of book stands scattered through the neighborhood. The stands were semipermanent structures, urban lean-tos supported by the tenements on one side, with counters and shelves improvised from discarded doors, window shutters, and stray planks of wood. Unlike the pushcarts, the book stands dealt mostly in new merchandise, books that were published by immigrants for immigrants, often produced in small local print shops.

Sometime in 1901, the same year the Rogarshevskys landed in New York, a skinny paperbound volume made its first appearance on the East Side book stands. The *Text Book for Cooking and Baking* by Hinde Amchanitzki was America's first Yiddish-language cookbook, a photograph of

the author, her wig neatly parted, gracing its cover. Very little is known
about the author's own immigration history, though in her foreword she
shares details from her culinary past. Amchanitzki's career as a professional
cook started in Europe and continued in America. Her recipes, she writes,
are based on forty-five years of experience working in both private homes
and restaurants, including an extended stint in a New York establishment
that catered to "the finest people." Amchanitzki's intended readers were
women much like Mrs. Rogarshevsky—seasoned homemakers trained in
one culinary tradition, now ready, in their own cautious way, to take on
another. Accordingly, the recipe index skips between the Old and New
World kitchens, with stuffed spleen, chopped chicken liver, and sponge
cake alternating with breakfast pancakes, tomato soup, and banana pie. A
third group of dishes, however, falls somewhere between the two kitchens, an amalgam of New World ingredients and Old World techniques.
Included in this category is Amchanitzki's recipe for cranberry strudel:

CRANBERRY STRUDEL

Take a quart of good cranberries, a half pound of sugar,
and a bit of water. Cook until thick and put aside to cool.
Take a glass of fat, a glass of sugar, 2 eggs, and stir together.
Add a glass of water and mix well. Take two glasses of
flour, two and a half teaspoons baking powder, mix them
together, and stir into batter. Take a sheet and grease it well.
Pour in half the batter and spread it evenly over the entire
sheet with a spoon. Spread the cranberries evenly over the
dough and pour the remaining dough over the cranberries,
covering them completely. Sprinkle sugar on top and bake
thoroughly. When done, let cool and cut into pieces. This is
a very good strudel.[13]

Early twentieth-century cookbooks brought news from the American kitchen to immigrants with limited access to the food habits of mainstream America. One place where contact was possible was the settlement house. The idea behind the settlement house, a British invention of the 1850s, was to bring together the educated and laboring classes for the benefit of both parties. It was always assumed, however, that the educated person had more to give, the laborer more to gain. America's first settlement house, the Neighborhood Guild (later known as the University Settlement), opened in 1887 on Eldridge Street on the Lower East Side, followed by Hull House in Chicago two years later. By 1910, the number of settlement houses in the United States had reached four hundred.

The settlement house aspired to elevate the working person to "a higher plane of feeling and citizenship." Most offered classes in literature, music, theater, and dancing, with kindergartens for the youngest children, clubs and gymnasiums for the older ones, and reading rooms for the adults. Some offered vocational training—more for women than men—providing instruction in millinery, sewing, and nursing. In immigrant-dense neighborhoods, where settlement houses assumed the job of Americanizing the foreign-born, there were also English-language classes, classes in civics, and American history. At the Educational Alliance, for example, immigrants could enroll in a lecture course that covered federal and state history, geography, government, and American customs and manners. Students who completed the course received copies of both the Constitution and Declaration of Independence printed in English and Yiddish.

Similar efforts to Americanize the immigrant took place in settlement cooking classes. The class curriculum was shaped by a relatively new approach to housework, known as domestic science, a movement that gained ground with American homemakers in the final decades of the nineteenth century. The women who led the domestic science charge set out to ennoble the homemaker's daily grind of cooking and cleaning by grounding it in scientific theory and method. They envisioned the home as a kind of domestic laboratory in which women applied their knowledge of chemistry, sanitation, dietetics, physiology, and economics to the everyday work of cooking and cleaning. To help spread their gospel, they established cooking schools in New York, Philadelphia, Boston, and Chicago, and domestic science programs in colleges and universities, including the Pratt Institute in Brooklyn and Columbia University in Manhattan. The Russian-American writer Anzia Yezierska was granted a scholarship to study at the Columbia School of Household Arts, but private cooking programs were generally beyond the means of working-class women. The domestic-science movement reached the working class through charitable institutions like churches, YMCAs, and settlement houses, where classes were offered for free or at prices scaled to the working person's budget.

The classes focused on the simplest and plainest American foods, beginning with a lesson on how to make toast and brew coffee in a freshly scoured coffee pot. Students learned how to properly boil oatmeal, rice, and potatoes, how to make pea soup, mutton stew, creamed

A cooking class for immigrant girls at the Educational Alliance. The girls here are learning how to make corn muffins.

codfish, biscuits, and gingerbread. Settlement houses that catered to Jews adapted the standard lesson plan so it conformed to Jewish dietary law, but only because they had no choice. The people who ran the settlement houses were Reform Jews, many from German families, who had immigrated to the United States in the mid-nineteenth century. In other words, oyster-eaters. In their desire to Americanize the immigrant, they would have preferred to dispense with kosher laws—and some, in fact, tried—but their students wouldn't allow it. As one settlement worker explained it, "There are some kosher laws that have to be followed, else the teaching would go no further than the classroom and would never show practical results."

Of the eight thousand East Siders that passed through the doors of the Educational Alliance each day, the vast majority came to take advantage of the free legal aid, use the public baths, or let their children loose on the rooftop garden. The cooking classes were only a modest success. The reason was simple: the Jewish homemaker already knew how to cook.

In 1916, the New York Board of Health issued a recipe booklet of cheap and nutritious foods intended for the East Side homemaker. Distributed by neighborhood settlement houses, *How to Feed the Family* was something of a flop among its intended audience. A reporter curious about the East Sider's reaction found out why. (The interview is transcribed in dialect, a common practice in period newspapers when dealing with working-class subjects.) The Board of Health, one woman explained,

> *ain't got no right to say what I should cook and how. Y'understand?*
> *Already when I was little I knew how oatmeal it should be cooked.*
> *You do it with a double boiler. I ain't got no use for peoples what*
> *teaches me how to cook things that a long time before I done better as*
> *what they did.*

Her neighbor concurred. "The East Side is the East," she told the reporter. "I make like my Grossmutter Selig and my mother Gefullte fish and stuffed *helzel* [poultry neck]. What I care for the Board of Health?"[14] But if the Board of Health failed to impress them, immigrant cooks felt the pressure to Americanize from other sources. The most persuasive were the cook's own children.

No single institution exerted more influence on the culinary lives of immigrant children than the American public school. Here, beginning in 1888, immigrant daughters were taught the fundamentals of American cookery in a then-experimental course based on the principles of domestic science. Declared a success by city educators, the experiment in "manual training" (the classes also gave instruction in sewing, housekeeping, and nursing) became a permanent fixture of the New York public schools. Over the next decades, it expanded and evolved along with the changing profile of the city. For poor students, classes in manual training opened up employment opportunities, but middle-class girls could benefit too. The classes taught them discipline, neatness, and organization, the qualities that would help them manage their own future households.

The program's original audience was native-born American girls, but the focus shifted as immigrants continued to descend on New York. To reach their foreign-born students, educators hit upon a novel teaching strategy. They replaced the conventional classroom with "model flats," simulated tenement apartments that mimicked the students' own tenement homes. In their stage-set kitchens, the girls learned how to maintain the highest sanitary standards, every dish and utensil neatly stowed in its rightful place. They were taught the importance of established mealtimes, the family sitting down together at a properly set table, the food "served" rather than "grabbed." Finally, they were tutored in the science of cooking with lessons on food chemistry, kitchen mechanics, and human physiology.

Model flats were the brainchild of Mabel Kittredge, a domestic scientist who worked with Lillian Wald at the Henry Street Settlement. The first model flat opened in 1902 in a tenement building in the heart of the Jewish ghetto. Before long, they were scattered through the tenement district, some housed in actual tenements, others in school buildings, including P.S. 7 on Hester Street. Miss Kittredge developed a housekeeping curriculum based on the model flats, which she compiled into a textbook. Today, *Practical Homemaking* provides a detailed picture of the public school cooking class circa 1914, when the book was published. The recipes in *Practical Homemaking*, hand-selected for the tenement population, were centered around three core ingredients: milk, cereals, and potatoes. Miss Kittredge saw little use for vegetables, with the exception of beans, the only form of plant life rich in "nutritive value." She was equally unimpressed by fruit, which, after all, was composed mostly of water. The immigrants' first cooking lesson was devoted to nature's most perfect food, milk, from which the girls were taught to make cocoa. Future lessons were devoted to white sauce, boiled cereals like oatmeal

Dieticians from local schools and settlement houses paid visits to the tenements. The dietician pictured is teaching immigrant women how to cook hot cereal in a double boiler.

and Wheatena, boiled potatoes, and cooked apples. Promoting the foods that Kittredge felt were best suited to the East Sider, the lessons were also designed to wean immigrants away from their less desirable culinary habits. For Jews, that meant forsaking their over-spiced pickles and delicatessen meats, while Italians were asked to cut back on their beloved macaroni and olive oil. Returning to their real-life tenement flats, the girls shared what they had learned, teaching their mothers how to poach eggs, or cook vegetables in boiling water rather than goose schmaltz. Teachers also made home visits to reinforce the lessons and monitor their students' progress. As one contemporary described it, the girls served as missionaries to their foreign-born parents, a role that the public schools exploited for all it was worth.

A second powerful influence on the food habits of immigrant children was the school lunchroom. Up until the twentieth century, most city kids returned from school each day for a home-cooked meal. That began to change as more and more women found work outside the home, leaving their kids to fend for themselves. With no one to feed them, the kids were given two or three pennies to buy lunch from a local pushcart or delicatessen. In 1908, a group of private citizens, alarmed by this new development, founded the New York School Lunch Committee, a charity that provided three-penny lunches to undernourished children. In place of pickles and candy—the typical pushcart meal—the committee provided hot soups and stews for two cents a serving, and one-penny treats like rice pudding or baked sweet potato. The school lunch committee lasted through World War I, but in 1920, responsibility for feeding the city's children shifted to the Board of Education. As it happens, the shift coincided with a groundswell of anti-immigrant thinking in the United States, which culminated in the Johnson Reed Act, a far-reaching immigrant quota system passed by Congress in 1924. Calls to Americanize the foreign-born reverberated through government offices and monopolized the editorial pages of the nation's leading newspapers. With so much attention on the immigrant threat, the Board of Education looked to the school lunchroom to Americanize the immigrant palate. Below is a typical school lunch menu circa 1920:

MONDAY: Cocoa, buttered roll, stewed corn, stewed prunes
TUESDAY: Cream of pea soup, peanut and cottage cheese
sandwich, Brown Betty with lemon sauce, fruit tapioca
WEDNESDAY: Vegetable soup, baked beans, vanilla corn-
starch with chocolate sauce
THURSDAY: Lima bean and tomato soup, buttered roll,
cream tapioca, rice pudding
FRIDAY: Cocoa, salmon sandwiches, sliced fruit, oatmeal
cookies[15]

To extend the lunchroom's influence, mothers were invited to eat with
their kids. During the meal, domestic-science teachers would point out
the benefits of the particular dishes served, urging them to prepare similar foods in their own homes. Across America, educators seized on the
lunchroom's educational possibilities, establishing similar programs of
their own. A domestic-science teacher named Emma Smedley summed
up the new awareness most succinctly. "No branch of the school activities," she wrote, "offers greater opportunity of fitting in with the Ameri-
canization plan than the school lunch."[16] The process was gradual, but
in the school lunchroom, kids of diverse backgrounds found a culinary
common ground, one tentative bite at a time.

In 1884, a new entry made its debut in Trow's New York Business Directory, ancestor to the modern-day Yellow Pages. Sandwiched between
"Deeds (Acknowledgement of)" and "Dental Equipment" now appeared
"Delicatessens." These specialized groceries had existed in New York
for at least thirty years, most of them clustered along First and Second
Avenues. Even so, 1884 was a kind of birthday for this immigrant food
shop, the year it captured the attention of the Trow's editors, asserting its
place in the city's food economy.

If the New York Tribune is correct, the city's first "delicatessen handler,"
or deli man, for short, was an immigrant named Paul Gabel, who landed
in New York in 1848, the year of revolution in Europe and the start of
the great German migration. (Gabel made a good living in America. By

1870, he had moved his store and his family to stately Brooklyn Heights, his fortune now worth $20,000, a substantial amount by the standards of the day.) Shops like Mr. Gabel's carried a limited stock of sausages, cheeses, and sweets, but as the century progressed, delicatessens added "made dishes" to their lineup of provisions—foods that were cooked and ready to eat, prepared by the owner's wife in a small kitchen behind the store. Hungry city-dwellers visiting their local delicatessen could choose among the following: meat pies, smoked beef shoulder, smoked tongue, smoked fowls, roast fowls, smoked, pickled, and salted herring, fresh ham, baked beans, potato salad, beet salad, cabbage, parsnip, and celery salads, in addition to all the usual wursts, breads, and cheeses. Though still in the hands of German New Yorkers, the delicatessens' clientele had now widened to include the city's growing population of Irish immigrants, along with native-born Americans. By the 1890s, delicatessens were "as common as bricks in a building"; the great majority, however, could be found on the Lower East Side. Rich New Yorkers, with their live-in servants and private cooks, had little real need for the delicatessen. But among the tenements, delicatessens assumed the role of a poor person's catering shop. Bachelors, shopgirls, boarders and lodgers, working mothers, people with little time for the kitchen, some with no kitchens at all, relied on the delicatessens to cook for them. Their services were particularly indispensable during the hot summer months, when firing up the kitchen stove turned the tenement apartment into a sweatbox.

In the first decades of the twentieth century, the "delicatessen habit" moved up the economic ladder and caught on among the middle class. With this development, a new tradition was born: the Sunday delicatessen supper, a meal composed of cooked foods, hot and ready to serve. Not only in New York, but across urban America, the delicatessen was now so thoroughly entrenched that it sparked an anti-delicatessen backlash. Domestic scientists, among other concerned Americans, blamed the delicatessen for an array of social maladies. A few links of sausage, a loaf of white bread, and a bottle of ketchup, the standard delicatessen meal, drives the workingman straight to the nearest saloon, these women argued. Along with intemperance—a source of growing apprehension in pre–World War I America—delicatessens were thought responsible for the nation's climbing divorce rate. "If fewer women depended on the delicatessen store," one expert argued, "there would be fewer broken homes."[17] Disgruntled husbands could be made manageable if their wives would only take the trouble to cook for them.

The history of the Jewish delicatessen follows a separate but roughly parallel track. The country's first Jewish delicatessens opened for business on the Lower East Side early in the 1850s. Established by German Jews, they specialized in smoked, brined, and spiced meats, much like their Gentile counterparts. They also carried myriad forms of herring, pumpernickel, and the standard assortment of German salads. The two stores even looked alike. The delicatessen's main staging area was a white marble counter, where the meats were displayed and sliced for the customer. The salads were arrayed in a row of stoneware crocks. What set the Jewish delicatessen apart was the total absence of any product derived from pigs. In its place, German Jews turned to geese. The following description is taken from an 1897 story that ran in the *New York Tribune*:

> *There are delicatessen shops in New York where roast fowl and sliced ham are unknown, where pigs' feet would not be tolerated, and where an order of venison would be given in vain. The Kosher delicatessen places of the crowded East Side, although in name like those in Sixth-ave., carry a stock of goods unlike those of any other place. There, in season, may be bought the various dainties made from goose meat. Among these are* Gansekleines, Gansegruben, *and fattened goose liver.* Gansekleines *is the name given to the small pieces of the dressed goose, like wings, feet, and neck, and* Gansegruben *are the pieces of the brown crackling from which the fat has been extracted. In some of these places they also prepare what is known as* Gesetztes Essen. *This consists of a mixture of barley and dried peas, which is prepared on Friday for consumption on Saturday when the pious Jews do no cooking.*[18]

Without a doubt, the Sabbath stew glistened with goose schmaltz.

Beyond these goose-based dainties, Jewish delicatessens sold kosher wursts and frankfurters, corned beef and corned tongue. In the early days, the cured meats were shipped over from Germany, but as the Jewish community settled in, it became more self-sufficient. During the 1870s, kosher sausage factories sprang up on the Lower East Side to supply the quickly multiplying number of Jewish delicatessens in New York, Philadelphia, Baltimore, and smaller cities as well. In 1872, a German butcher named Isaac Gellis produced some of America's first domestic kosher frankfurters in his sausage factory at 37 Essex Street. As the company was passed down from father to son to grandson, it grew into an empire, with delicatessen restaurants selling nothing but the Gellis brand scattered through Manhattan. One of them, Fine & Schapiro, can still be found on the Upper West Side of Manhattan. Moses Zimmerman, also from Germany, was another early sausage-maker. His factory on East Houston Street opened in 1877, producing bolognas, frankfurters, wienerwursts, corned beef, and corned tongue, along with kosher cooking fat.

In the 1880s, as migration patterns shifted and large numbers of Eastern European Jews sailed for America, they discovered the delicatessen in cities like New York, Philadelphia, and Chicago. The great majority had never seen one before. Older immigrants looked on the delicatessen with suspicion (it was "too spicy" and "too fancy"), while their children were intoxicated by its distinct perfume, a blend of boiled beef, garlic, pepper, and vinegar. Mondays through Fridays, they rushed from school to the local deli for a lunch of pickles and halvah. On Saturdays, the delicatessen was closed for the Sabbath, but it opened again Saturday evenings at sundown. This was a moment that ghetto kids looked forward to with crazed anticipation, famously captured by Alfred Kazin in his food-rich memoir, *A Walker in the City.* Saturdays at twilight, Kazin writes, neighborhood kids haunted the local delicatessen, waiting for it to reopen. As soon as it did, the kids raced in, "panting for the hot dogs sizzling on the gas plate just inside the window. The look of that blackened empty gas plate had driven us wild all through the wearisome Sabbath day. And now, as the electric sign blazed up again, lighting up the words Jewish National Delicatessen, it was as if we had entered our rightful heritage."[19]

The irony here is that the delicatessen was not, in fact, a pillar of

Jewish food culture, at least not for the Russians, or the Poles, or the Litvaks, but the Jews declared it one all the same. If their mothers disapproved—and most mothers did—for Jewish kids, the delicatessen was like a second home, part lunchroom, part urban clubhouse, and at night, an after-hours meeting place for ghetto sweethearts. With their limited pocket money, East Side kids were confined to the cheapest items on the delicatessen menu: a frankfurter with yellow mustard or a salami sandwich. The big-ticket item was a plate of sliced deli meat served with a tub of pickles. The most aristocratic option of all was the "mixed plate": a combination of pastrami, corned beef, and tongue.

Inch by inch, their kids leading the way, the new Jewish immigrants developed a taste for the cured meats of their German brothers and sisters. Those with an entrepreneurial bent looked to the delicatessen as a business opportunity and opened stores of their own. Samuel Chotzinoff, a Russian immigrant and future concert pianist, remembers exactly what that entailed. The Chotzinoff family arrived in New York in the late 1890s when Samuel was around eight years old. A few years later, when his mother decided to open a delicatessen, she paid a visit to one of the local sausage manufacturers. In keeping with East Side custom, the Mandelbaum Sausage factory offered her a kind of delicatessen start-up package. It included fixtures for the store (paid for on an installment plan) and three months of credit toward supplies. The sausage people even taught her how to cut meat into the thinnest possible slices, the delicatessen's key to financial success. A seltzer company lent Mrs. Chotzinoff a soda-water fountain for making syrup drinks. The store kitchen was a backroom with a three-burner stove, where she cooked her own corned beef in a tin clothes boiler. The entire operation cost her only $150 upfront, money that she borrowed from a well-off landsman.

Looking beyond the delicatessen, the Jewish East Side was home to a staggering variety of eating places. Neighborhood restaurants catered to every nameable niche and subniche of the local population—geographic, economic, professional, and even political—attracting a highly specialized clientele of like-minded diners. Every national group had its corresponding restaurant. The more modest establishments were located in tenement apartments temporarily converted into public eating spaces. Blurring the line between home and business, the private restaurant offered diners a truly Old World eating experience: a home-cooked meal prepared in the style of a particular region or city back in Europe. A visitor to the East Side in 1919, who discovered these private restaurants, explains how they worked:

> A great many of the emigrants from Russia and Rumania, even after years of alienation, have an intense craving for the dishes of their native province. They cannot assimilate the American cuisine, even though they accept its citizenship. It is, therefore, the practice of the inhabitants of particular province to convert her front parlor (usually located on the ground floor of a tenement) into a miniature dining room, where she caters to a limited number of her home-town folk. Her shingle announces the name of her province, such as "Pinsker," "Dwinsker," "Minsker," "Saraslover," "Bialystoker," etc., as the case may be. Here the aliens meet their friends from the Old Country and lose their homesickness in the midst of familiar faces and dialects and in the odors from the kitchen, which evoke for them images for their home and surroundings.[20]

Parlor restaurants answered the needs of the working person, but the ghetto also provided for the local population of well-off merchants, factory owners, lawyers, doctors, and real estate barons. By the turn of the century, a half dozen glittering eating-places had opened on the Lower East Side, which catered to the downtown aristocracy. Most of them were in the hands of Romanian Jews, the self-proclaimed bon vivants of the ghetto.

The Romanian quarter of the Lower East Side began at Grand Street and continued north until Houston Street. It was bounded on the west by the Bowery, the border between the Jewish ghetto and Little Italy, and by Clinton Street to the east, the thoroughfare that separated the Romanians from the Poles. The streets within this square quarter-mile were unusually dense with pastry shops, cafés, delicatessens, and restaurants,

the most opulent eateries south of 14th Street. Dining rooms were decorated in the sinuous Art Nouveau style, a raised platform at one end for
the house orchestra, the tables arrayed along a well-polished dance floor.
Sunday nights, when ghetto restaurants were at their busiest, the dance
floors were crowded with ample-bodied Jewish women, the grand dames
of the Lower East Side, decked out in their finest gowns and sparkliest
diamonds.

The deluxe surroundings belied the earthy, garlic-laced cuisine typical of the Romanian rathskeller. The following account of Perlman's
Rumanian Rathskellar at 158 East Houston Street comes from a 1930
restaurant guide:

> *The food for the most part is invariably unspellable and wholly deli-*
> *cious. Sweetbreads such as you never encountered before; smoked goose*
> *pastrami, aromatic salami, chicken livers, chopped fine and sprinkled*
> *with chopped onions; Wiener schnitzel; pickled tomatoes and pickled*
> *peppers; sweet-and-sour tongue; and huge black radishes. Because it's*
> *so good, you eat and eat until your head swims, drinking seltzer to*
> *help it along.*[21]

The Romanian restaurants were also known for their "broilings," or
grilled strip steaks, and for their *carnitzi*, sausages that were so pungent
they seemed one part ground meat to one part garlic.

Romanians shared East Houston Street with Hungarians, and
together the two groups transformed a generous chunk of the Lower
East Side into New York's leading café district. Where Russian Jews were
devoted tea drinkers, the Hungarians had acquired a love for coffee, a
habit learned from the Ottoman Turks. (Along with Austria, Hungary
was part of the vast territory claimed by the Ottomans between 1544
and 1699.) Settling in the United States in the late nineteenth century,
the Hungarians brought their coffee habit with them, establishing scores
of coffee houses in immigrant enclaves. Visitors to the East Side counted
at least one café on nearly every block of the Hungarian quarter, while
some streets had four or five. Coffee on the East Side was served in the

European style, with a small pot of cream and a tumbler of water, a symbolic gesture of hospitality. That was for patrons who asked for their coffee *schwartzen*. Coffee with milk was served in a glass. Whichever style, Hungarian coffee was often consumed with pastry, maybe a slice of strudel, apple or poppy seed, or a plate of *kiperln*, the crescent-shaped cookies that we know as rugelach.

After dark, well-heeled New Yorkers descended on the cafés for a night of "slumming," a term coined in the nineteenth century. For the uptown city-dweller, slumming on the Lower East Side was both an opportunity for cultural enrichment, like a visit to the museum, and a form of ribald entertainment. The adventure began as the uptowner crossed 14th Street and entered the foreign quarter, seeking immigrant cafés with olive-skinned waitresses, gypsy violinists, and fiery (to the uptown palate) Hungarian cooking. A favorite destination was Little Hungary, a haunt of Theodore Roosevelt during his term as New York police commissioner. Below, a 1903 guide to the East Side cafés deciphers the menu at Little Hungary for the bewildered uptown diner. First among the entrées is, of course *"Szekelye Gulyas,"* a sharp-seasoned ragout of veal and pork, with sauerkraut. Then there are such things as:

> *Lammporkolt mit Eiergeste*—a goulash of lamb
> *Peishel mit Nockerln*—a goulash of lung
> *Wiener Backhendle*—fried chicken, breaded
> *Kas-Fleckerl*—vermicelli with grated cheese
> *Zigeuner-Auflauf*— vermicelli with prune jelly
> *Palacsinken*—a sort of French pancake
> *Kaiserschmarren*—a German pancake cut into small pieces
> while baking, and mixed with seeded raisins
> *Strumpfbandle*—noodles with cinnamon and sugar
> Among the most noted pastries are *Apfel-Strudel, Mohn,* and
> *Nuss-Kipferl*[22]

It didn't seem to matter to the uptown patrons that the café crowd was made up of fellow slummers. High on slivovitz, they tumbled into their waiting carriages and bounced homeward, their taste buds still reeling from the onslaught of garlic and paprika.

The Russian quarter of the Lower East Side hosted its own café network, only here the action unfolded around steaming glasses of amber-colored tea. In his drinking habits, the Russian Jew was the inverse of the Irishman. The Irishman drank his tea at home, but socialized over whiskey in the East Side saloons. The Jew, by contrast, consumed his alcohol around the family table while tea was his drink of public fellowship. Café tea was brewed in samovars and served in glass tumblers with a thick slice of lemon and a lump of sugar that the drinker clamped between his front teeth. The hot liquid was then sucked through the sugar with a loud, slurping sound. In the process of drinking, a few tablespoonfuls always splashed over the edge of the tumbler into the saucer beneath. When the glass was empty, the drinker raised the saucer to his lips and drained that as well.

In Little Hungary, music was part and parcel of the café atmosphere. In the Russian quarter, music was replaced by the sound of talk—feverish, theatrical, and at times contentious. The food and drink were secondary. For this reason, the cafés came to be known as *kibitzarnia*, from the Yiddish verb *kibetzn*. Roughly translated, to kibitz is to banter, in a sometimes mocking or intrusive way. The Yiddish writer Sholem Aleichem, who was born in Russia in 1859 and lived for a short time on the Lower East Side, explains that to kibitz "is to engage in repartee of a special sort, to needle someone, tickle him in the ribs, pull his leg, gnaw at his vitals, sprinkle salt on his wounds, give him the kiss of death, and all with a sweet smile, with a flash of rapier-like wit, with whimsy and humor . . ." In the *kibitzarnia*, he continues, the customer orders a cup of tea and a bite to eat, in prelude to the real action. Now the kibitzing starts:

> *The barbed compliments fly between the tables. Racy stories and witticisms are passed around, each calculated to step on someone's toes where the shoe pinches most . . . The kibitzarnia, dear reader, is a sort of free Gehenna [hell] where people rake each other over the coals, a steam bath where they beat each other with bundles of twigs until the*

blood spurts. Here opinions are formed, reputations are made and destroyed, careers decided.[23]

During the day, the café doubled as a conventional working person's restaurant serving traditional Russian fare: bean soup, borscht, kasha varnishkes, and herring in all its forms. At Leavitt's Café on Division Street, stomping ground of the East Side literary set, patrons could order a plate of chopped chicken liver for a nickel, or meatballs with farfel for 15 cents. After the dinner hour, the café assumed its nighttime role as the local debate club/lecture hall/classroom/salon, where the talking continued unabated until two or three in the morning. In the East Side hierarchy of daily necessities, good conversation trumped a good night's sleep.

A café existed for patrons of every political conviction. All of the "-ists" were represented: Socialists, Marxists, Zionists, Bundists, anarchists, and even the lonely capitalist, odd man out among Russian Jews, could find a sympathetic audience in one café or another. And not just men, but women too, were at home in the politically charged café environment. To the uptown observer, the sight of these "unwomanly women," sitting in mixed company and denouncing this or that government, came as a shock. To the café regulars, hungry for revolution on any front—political, artistic, or social—it exemplified the new world to come. Revolution, however, was for the young. . . . The older generation had their own establishments, where the talk centered on spiritual matters, and men in derbies (the café patron never removed his hat) disputed esoteric points of Talmud over a glass of tea and a slice of strudel. This is where we may have found the pious Mr. Rogarshevsky, president of his synagogue, embroiled in religious debate with his fellow congregants. The café also served the ordinary working people— the factory hands, shopkeepers, and peddlers who were more concerned with earning a living than with Nietzsche or Marx. This last group, repairing to the café at the close of the workday, sucked down glass after glass of hot tea—ten, twelve, fifteen glasses in succession—to soothe their throats, raw from a day of shouting.

Another expression of culinary specialization could be found in the East Side knish parlors, restaurants dispensing that one item and not much else. Starchy and filling, the knish, or *knysz*, was Russian peasant food, a rolled pastry traditionally filled with kasha. Because it was portable, it could be carried into the fields for a calorie-dense midday meal. Adopted by the Jews, the knish was transplanted to America, where it became the quintessential East Side street snack. Hot knishes were initially sold from carts that resembled tin bedroom dressers but were actually coal-burning ovens on wheels. The knishes were stored in the warming drawers. Like other East Side street foods—the bagel included—the knish eventually moved inside to a proper shop, the knishery.

The Jews made two basic types of knishes, *milchich* and *fleishich*, dairy and meat. The dairy knish was filled with pot cheese, the meat knish with liver. Knishes filled with kasha, potato, and sauerkraut could go either way, their status determined by the type of shortening used in the dough, butter or schmaltz. Deep into the 1920s, East Side knish-makers still followed the traditional strudel-like blueprint, stretching their dough, slathering it with one or another filling, then rolling it up like a carpet. After baking, the baton-shaped pastry was cut into sections. But this represented only one possible configuration. When uptowners discovered the knish sometime around World War I, they were baffled by it. A visitor to the East Side in 1919, on a tour of local eating spots, responded to the knish with typical befuddlement:

> *Another institution which is part of the multifarious life of the lower East Side is the "knishe" restaurant. The "knishe" is a singular composition. One may look in all the cook books and culinary annals of all times for the recipe of a "knishe," but his efforts will be futile. Its sole habitat is the East Side.*[24]

And so began the mythological association between this Slavic pastry and New York City, birthplace of the knish in the American culinary imagination.

Everything about the knish was so well-suited to the mode of life on the Lower East Side that it seemed to have sprung from the asphalt. Its portability was one of its major assets. Another was its price. What other food could deliver so much satisfaction for only three cents? At midday, it was a cheap and filling lunch for the sweatshop worker. At night, theatergoers devoured a quick knish at intermission or stopped by the local knishery for an after-show snack. In fact, an important connection developed between knishes and theater that helped establish a place for the knish in the local ecosystem. The home of the Yiddish theater in the early twentieth century was Second Avenue, the playhouses concentrated between 14th and Houston streets. That same strip came to be known as "knish alley" in recognition of the many knish joints that had sprung up within a few blocks' radius. Knish parlors followed theaters the same way that pilot fish follow sharks. Wherever a theater opened, a knishery followed. In 1910, a Romanian immigrant named Joseph Berger opened a knish restaurant at 137 East Houston Street, directly next door to the Houston Hippodrome, a Yiddish vaudeville house that also showed moving pictures. The restaurant was named for Berger's cousin and former partner, Yonah Schimmel, the knish vendor who had started the business two decades earlier with a pushcart on Coney Island. Berger's son, Arthur, took over from his father in 1924, and continued selling knishes for the next fifty years. Eventually, the business was sold to outside investors. The Hippodrome building is still a movie theater, now a five-screen multiplex. When the shows let out, hungry theatergoers walk the same thirty feet to Yonah Schimmel's, functioning time capsule and last of the East Side knisheries.

One measure of wealth among East Side Jews was how much meat a person could afford. Because they came from a meat-scarce society, its sudden availability in America represented the unlimited bounty of their adopted home, and Jews aspired to eat as much of it as possible. This fixation on meat helps explain the exalted place of the delicatessen in the life of the ghetto. But shortly after the turn of the century, a new type of eating place appeared on the East Side, which served no meat at all: the dairy restaurant. Here, with the exception of fish, the kitchen was strictly vegetarian, concentrating on foods made from grain, vegetables, milk, and eggs. On the face of it, the dairy restaurant was a natural outgrowth of the Jewish dietary law that forbids the mixing of meat and milk. On closer inspection, however, its appearance in New York around 1900 was a product of culinary forces that extended beyond the ghetto.

First, the East Side dairy restaurant was part of a growing interest in vegetarian dining that had recently taken hold of New York. The city's first vegetarian restaurant opened in 1895 and more followed, providing patrons with a meatless menu of salads, nut-butter sandwiches, omelets, vegetable cutlets, and dairy dishes like berries and cream. American vegetarians came to their dietary views by way of religion. One early proponent was the Reverend Sylvester Graham (advocate for whole-grain bread) who helped found the American Vegetarian Society in 1850. Another was John Harvey Kellogg, a Seventh-Day Adventist and culinary inventor responsible for the creation of corn flakes. At his sanatorium in Battle Creek, Michigan, Kellogg also experimented with faux meat compounds made primarily from gluten and nuts, which became a staple of the East Side dairy menu.

The appearance of the first Jewish dairy restaurants coincided with a culinary crisis on the Lower East Side, which centered on the high cost of kosher meat. In the spring of 1902, a sudden jump in the price of kosher beef uncorked the pent-up outrage of East Side housewives. The women organized a neighborhood-wide boycott for the morning of May 15, with picketers stationed in front of every neighborhood butcher shop. Patrons who crossed the picket line had their purchases seized and doused with kerosene. At eleven a.m., a group of women and boys marched down Orchard Street and smashed the windows of every butcher en route, including the basement shop at number 97. Police who tried to stop the women became the target of their anger. The demonstrators pounced on the officers and wrestled them to the ground or pelted them with garbage. That night, five hundred women assembled at an East Side meeting hall. As the surrounding streets filled with angry supporters, tensions escalated between the crowd and the police. The inevitable fight broke out, and within the hour the neighborhood was engulfed in violence. The rioting subsided by the following afternoon, but the meat troubles continued for another decade, sparking boycotts and protests, though nothing on the scale of 1902.

The East Side's first dairy restaurants, born in the midst of the kosher-meat crisis, were shoestring operations, the menu limited to a handful of traditional dishes like blintzes, kasha, and herring. By the 1940s, however, this working person's lunchroom had evolved into a more ambitious enterprise. The most ambitious of all was Ratner's, which had originally opened in 1905 in a cramped storefront on Pitt Street. In 1918, the restaurant moved to its new home on Delancey Street, right next door to the Loew's Delancey, then a neighborhood vaudeville theater. In 1928, the Loew's Delancey became the Loew's Commodore, one of the new and fantastically ornate movie palaces that had begun to appear in the city. That same year, Ratner's received its own renovation at a cost of $150,000, transforming the old-time lunchroom into the "East Side's premier dining place." In its more elegant guise, its menu blossomed, and by 1940 covered a vast gastronomic territory ranging from the traditional herring salad to asparagus on toast to caviar sandwiches, among the most expensive items on the menu. But the most creative dishes to emerge from the dairy restaurant were their counterfeit meats. In place of actual beef or chicken or lamb, the dairy restaurants served meat substitutes that craftily mimicked the original. There was vegetarian stuffed turkey neck, chicken giblet fricassee, or chopped liver, all traditional Jewish foods. Diners with more assimilated taste could have vegetarian lamb chops or meatless veal cutlet. All of these foods were grouped under a section of the menu labeled "Roasts." Under the same heading was a selection of the faux meat products manufactured by Kellogg at his Michigan plant. The most popular was Protose steak, which the dairy restaurants served with fried onions or mushroom gravy. Here's a classic recipe for vegetarian chopped liver, with the "livery" taste surprisingly coming from the canned peas:

Lillian Chanales's Vegetarian Chopped Liver

3 medium-sized onions, chopped
3 tablespoons vegetable oil
I large can sweet peas, drained
I ½ cups chopped walnuts
2 hard-boiled eggs, chopped

Sauté the onions in the oil until they are soft and golden.
Mash peas with the back of a fork. Combine onion and
peas with remaining ingredients and chop by hand until you
have the desired consistency. If you like, you can use a food
processor, but be careful not to over-process. Season with
salt and a generous dose of freshly ground black pepper.[25]

The East Side's vast network of food purveyors satisfied the diverse
culinary needs of the local population with a thoroughness that was
unmatched in most other neighborhoods. People like the Rogarshevskys
had no reason to cross 14th Street to buy their horseradish or kosher meat
or to find a congenial café or a restaurant that met their religious standards.
As East Siders began to disperse, the food merchants followed. Kosher
butcher shops opened in Brooklyn, the Bronx, and along upper Broadway,
in addition to Jewish bakeries and delicatessens. Dairy restaurants began to
appear in midtown to feed the Jewish garment workers, and more opened
on the Upper West Side. But still, former East Siders returned to the old
neighborhood to shop from the downtown merchants and patronize the
restaurants. Immigrants who had moved to Brooklyn or the Bronx or
Upper Manhattan made Sunday trips to the Lower East Side to flex their
bargaining muscles at the pushcart market and buy a smoked whitefish
at Russ & Daughters, the appetizing store on Houston Street. Before the

holidays, they converged on the East Side to buy their matzoh, kosher wine, and dried fruit. When the shopping was done, they went for lunch at Ratner's or Rappaport's, another of the East Side's dairy restaurants.

Accounts of these food-inspired trips to the Lower East Side appear regularly in immigrant memoirs and immigrant fiction as well. A Fannie Hurst story called "In Memoriam" follows the tribulations of Mrs. Meyerberg, a lonely Fifth Avenue matron who returns—by chauffeured limousine—to her former tenement kitchen. Flooded with memories, Mrs. Meyerberg is moved by a sudden impulse to assume her place behind the tenement stove, and she does, but the experience proves too much for her. In typical Fannie Hurst fashion, the matron literally dies of joy. Anzia Yezierska's East Side heroine, Hannah Brieneh, makes a similar voyage. Now an old woman, residing in relative splendor on Riverside Drive, Hannah Brieneh is bereft, a living soul trapped in a mausoleum. The answer to her existential crisis is a trip to the pushcart market. "In a fit of rebellion," she rides downtown, buys a new marketing basket, and heads for the fish stand. The downtown foray is like a splash of cold water for the withering Hannah Brieneh, who returns in triumph, filling the lifeless apartment with the homey smells of garlic and herring.

The subway ride from the tenements to uptown New York proved more disruptive to immigrant food ways than the initial journey to America. Comfortably middle-class, the uptown Jew could eat like royalty, meat three times a day, unlimited quantities of soft white bread, pastry and tea to fill the gap between lunch and dinner. But uptown living came with unexpected constraints. Uptown Jews were plagued by a new and irksome self-consciousness that complicated mealtimes. Americanized children badgered their immigrant parents to give up the foods they had always relished. If the uptown Jew had a craving for brisket and sauerkraut, the aromas of these dishes cooking on the stove wafted through the apartment building and neighbors complained. The once-beloved organ meats became tokens of poverty, and uptown homemakers had to sneak them into the kitchen like contraband on the servant's day off. What a pleasure, then, to escape to an East Side restaurant for a plate of chopped herring and a basket of onion rolls.

CHAPTER FIVE

The Baldizzi Family

"Whoever forsakes the old way for the new knows what he is losing but not what he will find."
—SICILIAN PROVERB

At the start of the twentieth century, 97 Orchard Street stood on the most densely populated square block of urban America, with 2,223 people, most of them Russian Jews, packed into roughly two acres. One hundred and eleven of them resided in the twenty apartments at 97 Orchard, the oldest building on the block.

By the 1930s, the same East Side neighborhood was a shadow of its former self. Many of the older tenements had been abandoned by their owners, who could no longer afford to pay the property taxes, and were now vacant shells. Others had been demolished or consumed by fire and never rebuilt. As a result, a neighborhood once defined by its extreme architectural density was now littered with empty lots. The tenements that survived the 1920s were languishing too, the victims of changing demographics. Immigration had slowed dramatically by the middle of the decade; old-time East Siders, those who had settled in the neighborhood before the war, had dispersed to the outer boroughs. The number of people living at 97 Orchard, for example, had shrunk from one hundred and eleven to roughly twenty-five, leaving one-third of the building's apartments completely empty. The East Side tenant shortage meant that neighborhood landlords—even the most conscientious—could no longer afford to maintain their properties, and many buildings fell into disrepair.

Built during the Civil War, years before New York had formulated a body of housing laws, 97 Orchard embodied a laissez-faire approach toward lodging for the working class. As the building passed from one owner to the next, it was gradually modernized. In 1905, 97 Orchard was equipped with indoor plumbing. A system of cast-iron pipes now branched into every apartment and connected to the kitchen sink, supplying tenants with cold running water. The same system allowed for indoor water closets. A second major overhaul came in the early 1920s, when the building was wired for electricity. Despite these efforts, 97 Orchard remained an architectural relic. As late as 1935, the four apartments on each floor were served by two communal toilets. None had bathtubs or any form of heat apart from the kitchen stove. Only one room in three had proper windows.

In the years following World War I, 97 Orchard was home to a mix of Irish, Romanians, Russians, Lithuanians, and Italians. Included in this last group were the Baldizzis, a family of Sicilian immigrants that had come to New York to share in the unlimited possibilities of the American economy. Their plans, however, were derailed by the stock market crash of 1929 and the resulting disappearance of millions of jobs.

For most of the nineteenth century, as Germans and then Irish streamed into the United States, the Italian population stayed at microscopic levels. The 1860 census counted only twelve thousand Italian-born immigrants in the entire country, a demographic speck. The great majority of these early settlers were Northern Italians from Genoa, the surrounding province of Liguria, and from Piedmont just to the north. The numbers began to climb in the boom decades after the Civil War as America turned to the work of rebuilding. The rush of postwar construction activity created more jobs than the country could fill with its own citizens. So, America turned to her neighbors overseas. With the encouragement of the United States government, work-hungry Italians—among other immigrant groups—stepped in to alleviate a desperate labor shortage. During the 1880s, fifty-five thousand Italians arrived in the United States, and just over three hundred thousand in the decade following. By the end of the century, more Italians were landing at Ellis Island than any other immigrant group, and the trend continued into the 1900s. The immigrants who belonged to this second wave were overwhelmingly from the southern provinces of Basilicata, Calabria, and Sicily.

Two features of the Italian migration distinguished it from other groups. First was the lopsided ratio of men to women from Italy. During the last two decades of the nineteenth century, that ratio was four to one, with men leading the way. Though the numbers balanced out some over time, they never reached an even fifty-fifty. Lone Italian men came to the United States to work in railroad construction, to build dams, dig canals, lay sewer systems, and pave the nation's roads, "pick and shovel" jobs. The Italian laborer was typically a man in the prime of his working life. Many had wives and children back in Italy, to whom they planned to return once they had saved enough of their American wages to go back home and purchase a farm or maybe start a business. The average length of the Italian's sojourn in America was seven years. Some Italians became long-distance commuters. They worked in America during the busy summer season and returned home for the slow winter months, when construction was put on hold.

The success of this international labor pool hinged on a figure known as the padrone, an immigrant himself who wore many hats. Part employment agent, part interpreter, part boardinghouse keeper, and part personal banker, the padrone supplied the new immigrant with much-needed services while robbing him of half his wages, and sometimes more. The padrone's headquarters were in America but his work began in Italy, scouring the countryside for prospective clients—dissatisfied field workers, in good health, who were willing to travel. This work was often delegated to an Italian-based partner, who worked on commission. The

padrone also formed relationships with American employers who kept
him apprised of their labor needs, so when an immigrant landed, the
padrone knew where to send him. In the cities, he kept boardinghouses
where his clients were compelled to lodge, charging extortionist rates for
a patch of floor to sleep on. Italians who were sent afield to lay railroad
tracks or dig reservoirs in the American hinterland were beholden to
the padrone for all their basic needs. Other men who worked on these
grand-scale building projects—Slavs, Hungarians, and even the occasional American—lived in camps established by their employer. They
slept in the company bunkhouse and ate together in the company mess
hall. The one group missing from this international community was the
Italians, who followed their padrones to all-Italian camps complete with
bunkhouses, a commissary for buying supplies, and a kitchen where the
men ate. These boardinghouses in the wilderness, catering to a captive
and hungry clientele, were another money-maker for the padrone. At
the same time, they answered one very important requirement for the
laborer: to eat like an Italian.

Where other groups consumed whatever stews, breads, and puddings
they were given, the Italian demanded foods from the homeland. Over
the course of one month, the typical laborer consumed:

> Bread 34.1 lbs
>
> Macaroni 19.3 lbs
>
> Rice .24 lbs
>
> Meat (sausage, corned beef, & codfish) 2.31 lbs
>
> Sardines 2–5 boxes
>
> Beans, peas and lentils 2.06 lbs
>
> Fatback (lard substitute) 5.13 lbs
>
> Tomatoes 2.13 cans
>
> Sugar 2.8 pounds
>
> Coffee .43 lbs[1]

The men purchased their supplies at the commissary or shanty store, a
grocery run by the padrone, where everything on the shelves was triple

its normal price. The men took turns behind the camp stove, a group of three or four preparing meals for the entire crew. Some camps ran their own bakeries, using commissary flour. Out west, a similar arrangement could be found among Chinese railroad workers. In their separate camps, faced with the unappetizing prospect of the company mess hall, the Chinese workers assumed the job of feeding themselves, the only possible way to procure food that they considered edible. For both groups of men, Chinese and Italian, cooking became a New World survival skill.

Many of the foods that issued from the communal pot would be familiar today. Various forms of lentil soup, macaroni and tomatoes, beans and macaroni, beans and salt pork, and beans with sausage. Next to the familiar were more uncommon preparations. One ingenious food was a kind of homemade bouillon cube, prepared by Italian workers in Newark, New Jersey, circa 1900. Using a beer vat as their mixing vessel, the men first pounded a large quantity of tomatoes. Next,

> they poured some cornmeal and flour into the vat and stirred until the stuff became a dough. The next step was to throw this on what bakers would call a molding trough and knead it, adding enough flour to make it a stiff pulp. The less said about the state of their hands the better, but that is a trivial matter. The mixture was molded into little pats about the same size of fishcake. These were placed on boards and taken to various roofs to dry.[2]

Come winter, when fresh tomatoes were no longer available, the cakes were dissolved in boiling water, each cake producing enough soup for six men.

A second defining feature of the Italian migration was poverty. After 1865, the great majority of Italian immigrants were poor southern fieldworkers. They arrived at Ellis Island, illiterate and unskilled, with, in 1901, an average life-savings of $8.79. Despite their farming background, most Italians settled in the large industrial cities. Here they found work as street cleaners, pavers, and ditch- and tunnel-diggers—the dirtiest, most dangerous jobs. Immigration officials bluntly referred to the Southern Italian as America's "worst immigrants," a judgment echoed in the daily papers. "Lazy," "ignorant," and "clannish" were just a few of the adjectives most commonly linked with Italians by the popular press. "Violent" was another oft-mentioned Italian characteristic. American newspapers kept a running tally of crimes committed by the Black Hand, an early name for Italian organized crime, paying special attention to any case that involved explosives. (Bombs were a fairly common means of extortion among gangsters of the period.) America's fascination with Italian gangsters helped reinforce the argument that Italians were violent by nature. Following this circular logic, Americans were convinced that "no foreigners with whom we have to deal, stab and murder on so slight provocation," a judgment offered by the *New York Times*.[3]

Among the lowest of this low-grade stock were the men and women who rejected honest work in favor of more shiftless occupations. One character was the Italian organ-grinder, a roving street performer with a hand organ suspended from a strap around his neck. The hand organ worked like an oversized music box, with a rotating cylinder inside it that turned by means of a crank. The more prosperous worked with an assistant, a trained monkey in a red vest and matching fez. The animal perched on his master's shoulder while the music played and collected pennies at the end of the number. The organ grinder's main patrons were the city's children.

Another dubious line of work was rag-picking, an urban occupation dominated by foreigners, beginning with the Germans in the 1850s. The Irish also turned to rag-picking, but in smaller numbers. By the 1880s, the industry had been passed down to the Italians, the country's newest immigrants. America's first career recyclers, rag-pickers made their livelihood by sifting through the city's garbage for reusable resources. The tools of the rag-picker's trade were a long pole with a hook at one end and a large sack slung across her chest. Her workday began before the city was fully awake, when the streets were still quiet. She made her rounds, moving from one trash barrel to the next, examining its contents with the help of her pole, and plucking from it whatever she found of value. Her most fruitful hunting grounds were the cities' wealthier neighborhoods, where the garbage was rife with discarded treasure—old shoes and boots, battered cooking pans, glass bottles, and the rags themselves, cream of the trash barrel. Once home, she emptied her sack onto the floor to survey the day's gleanings. Each type of article was sorted into its own box, one for paper, one for leather goods, one for metal, one for glass, and so on. The bones were put into a large kettle and boiled clean. The rags were rinsed and hung up to dry.

The next stop for the sorted garbage was the junk dealer, a refuse middleman who paid the rag-picker a set sum by weight for each material, then turned around and sold it, at a profit, to assorted manufacturers. Old shoes and boots were retooled to look like new or shredded to a pulp, an ingredient used in the manufacture of waterproof tarps. Paper was sold to local publishers, who turned it into newsprint for the morning papers. Bottles were reused or melted. Bones from the family dinner table were turned into umbrella handles, snuff boxes, buttons, and toothbrushes. Rags, which fetched more per pound than any other item, went to make the era's finest writing paper.

Middle-class America declared the rag-picker too lazy for "real work," or accused her of ulterior motives. All of that innocent rummaging was, they believed, a cover for her real purpose—casing the best homes in the city for future burglaries. The organ grinder was likewise seen as a threat to public welfare, a nuisance at best, but at worst a common street thug, his stiletto tucked in his boot. Bootblacks, chestnut vendors, and fruit peddlers all belonged to the same itinerant class and all were suspect.

Non-Italians found proof of these immigrants' lowly character in the foods they ate: stale bread, macaroni with oil, and, if they were lucky, a handful of common garden weeds. No other immigrant diet was as meager. For his nourishment, the Italian fruit peddler relied on the bruised and moldy fruit that was too far gone to sell, even by East Side standards. Organ grinders, because of their aversion to work, subsisted on a diet "so scanty that had they not been accustomed to the severest deprivation from infancy, their system would refuse to be nourished by food that an Irish navvy would shrink from with abhorrence."[4] Their spare diet acted as an impediment that kept the immigrant from rising in the world. A typical lunch for the Italian laborer, a piece of bread and cup of water, was no meal for a working man. American employers, who could choose from an international pool of workers, came to regard the Italian as second-rate. "They are active, but not hardy or strong as the average man"—a rung below the Hungarian or the Slav. The main reason: "they eat too little."[5]

But no diet was more reviled than the rag-picker's—a hodgepodge of bread crusts, vegetable trimmings, bones, and meat scraps plucked from middle-class trash bins. One particularly desperate class of rag-pickers scavenged only for food, eating some of what they gathered and selling the rest for profit. Italian women from Mulberry, Baxter, and Crosby streets, the food scavengers targeted the city's markets, grocers, fruit stands, butchers, and fishmongers. An 1883 newspaper story from the *New York Times* describes how they operated:

> *Partially decayed potatoes, onions, carrots, apples, oranges, bananas, and pineapples are the principal finds in the mess of garbage that is overhauled. The greatest prize to the garbage-searching old hag is a mess of the outside leaves of cabbage that are torn off before the odorous vegetable is displayed for sale on the stands. The rescued stuff—cabbage leaves, onions, bananas, oranges, &c.—is dumped into a filthy bit of sacking, and the whole carted, as soon as a day's labor is concluded, to the miserable quarters which the old hag is forced to call home. Here a sorting process is gone through with. If the husband or son is sufficiently endowed with this world's goods to be the proprietor of a fruit-stand, everything that may possibly be sold for no matter how small a sum is transferred to him. The remainder is subdivided. The cabbage leaves which are fresh are sold for use in the cheap restaurants to be served with corned beef. Such as will not serve for that purpose is stripped of decayed portions and used as the body of the poor Italian's favorite dish—cabbage soup. In the composition of this dish, often for days at a time the only food save stale bread, which a family has to dine upon, are mixed the portions of the potatoes, carrots, and other vegetables which are not absolutely rotten. The tomatoes*

are used in making a gravy for the macaroni, this delicacy being se-
cured in exchange for decayed fruits and vegetables at the groceries or
restaurants of the Italian quarter. This process of exchange is carried
on quite extensively as the store-keepers prefer it, and find an addi-
tion to their meager profits in the system of barter.[6]

Despite the revulsion of middle-class America, the rag-picker's harvest provided her and her family with a windfall of edible wealth. American queasiness over "rescued food" was a luxury that the struggling immigrant could easily overlook. The heaps of discarded food, some of it perfectly good, which materialized each day in city trash bins, must have left the rag-picker gaping in wonderment. On the one hand, the rag-picker's lack of skills, education, and English left her consigned to the outer fringes of American society. Still, she was able to make a living off America's leftovers. American abundance was so staggering that the garbage that accumulated daily in cities like New York could support a shadow system of food distribution operated largely by immigrants. The rag-picker was a key player in this shadow economy, redistributing her daily harvest to peddlers, restaurants, and neighborhood groceries. In her own kitchen, the rag-picker's culinary gleanings formed the basis of a limited but nourishing diet. (Sanitary inspectors were often surprised by the rag-picker's good health.) Even more surprising, the rag-picker cook was determined to both nourish and delight, bartering for macaroni—a luxury food in the immigrant diet—while turning her rescued fruit into jellies and marmalade.

Expressions of anti-Italian bias continued until the start of World War I, when the nation shifted focus to fighting the Germans. German-Americans, once regarded as model immigrants, were now considered a threat to national security. Suspected of loyalty to the Fatherland, they were declared "enemy aliens" by President Woodrow Wilson and subject to a string of government restrictions. Thousands were arrested or interned. In the anxious years after the war, animosity directed at the Germans spread to other foreigners, placing immigrants and immigration at the center of a national debate. Though many of the old fears persisted, the new nativists turned to the faux science of race studies, a potent blend of anthropology, biology, and eugenics. American prosperity, they argued, rested on the superior mental traits of the Anglo-Saxon, attributes that were passed down from parent to offspring in much the same way as eye color. Alarmed by the recent influx of Southern and Eastern Europeans, the nativists claimed that decades of unchecked immigration had compromised the greatness of America, and the danger would continue as long as the gates stayed open. If they did, the outcome was assured: race suicide. By this reckoning, the settlement workers and schoolteachers who had worked so hard to Americanize the foreign-born were hopelessly misguided. American greatness could not be taught. It was literally in "the blood."

Among the leading voices of the new nativists was a New Yorker named Madison Grant, a lawyer and amateur zoologist who helped found the Bronx Zoo. Alongside his interest in wildlife, Grant developed a taste for politics, chiefly in the field of immigration policy. He encountered like-minded thinkers in such organizations as the Immigration Restriction League and the American Defense Society, a group originally founded to protect America from German aggression. Once the armistice was signed, the group shifted focus to a new enemy, the immigrant. Grant published his manifesto, *The Passing of a Great Race*, in 1916. The book never sold very well, but the ideas laid down by Grant filtered into Washington. Here, they became the quasi-scientific basis for a series of anti-immigration laws culminating in the 1924 Johnson-Reed Act, the most stringent immigrant quota system in United States history. After 1924, the total number of immigrants allowed to enter the United States each year became one hundred and fifty thousand, a minuscule number compared to earlier times. What's more, Johnson-Reed was specific about who those immigrants could be. America was now prepared to admit two percent of each foreign population living in the United States as of 1890, a year strategically chosen to make room for Western Europeans while shutting out less desirable types—Eastern Europeans, Italians, and Jews.

More than other groups, Italians arrived in this country with the firm knowledge that they were unwanted. In the workplace, Italians were paid less than other ethnicities, or denied jobs entirely. Landlords with a no-Italians policy denied them housing. The fact that few spoke English offered little protection against ethnic slurs, sources of the deepest humiliation for the transplanted Italian. The immigrant soon discovered that words like *dago* and *ginny* were accepted features of American speech, and not only in the streets and the schoolyards. "The Rights of the Dago" and "Big Dago Riot at Castle Gate" were the kinds of headlines Americans could expect to find in their morning paper. The quota laws effectively made anti-Italian discrimination the official policy of the United States government.

In the hostile environment first encountered by Italians, food took on new meanings and new powers. The many forms of discrimination leveled at Italians encouraged immigrants to seal themselves off, culturally speaking, from the rest of America. This circling of the wagons, a response typical of many persecuted people, was interpreted by Americans as Italian "clannishness," an unwelcome trait in any immigrant but especially so for an already suspect group—poor, uneducated, and Catholic. The metaphorical walls built up by Italian-Americans served a double purpose. On the one hand, they protected the immigrant from outside menace, both real and invented. On the other, they carved out a space where Italians could carry on with their native traditions in relative peace, away from American disapproval. As it happens, the traditions they seemed most devoted to were those connected with food. Certainly, culinary continuity was important to other foreign groups, but Italian-Americans were bonded to their gastronomic heritage with an intensity unknown to Russians, Germans, or Irish, and went to great lengths to protect it. The Jews had their religion; the Germans had their poets, their composers, and their beer; and the Irish had their politics. The Italians arrived with a strong musical tradition; they also had their faith. But food was their cultural touchstone, their way of defying the critics, of tolerating the slurs and all of the other injustices. It was their way of being Italian.

Harsh critics of Italian eating habits, Americans tried through various means to reform the immigrant cook. The Italians were unmoved. Despite the cooking classes and public school lectures, and despite the persistent advice of visiting nurses and settlement workers, the immigrants' belief in the superiority of their native foods was unwavering. Respect for the skills of the Italian cook, the goodness that she could extract from her raw materials, was one thing, but the immigrants were equally devoted to the materials themselves. For them, good Italian cooking was made from foods that grew from the Italian soil, and they used imported ingredients whenever possible. Olives and olive oil, anchovies, jarred peppers, dried mushrooms, artichokes cured in salt, canned tomatoes and tomato paste, vinegar, oregano, garlic, a variety of cured meats and cheeses, and, above all, pasta, were some of the products found in the Italian groceries that served America's many Little Italys. Here, Italian homemakers, working women of limited means, could stock their pantries with native foods, despite the daunting cost of imported goods. The financial sacrifice was proof of the Italian's dedication. A 1903 newspaper story describing the Italian grocery for readers who had never seen one took note of the Italians' food priorities:

> No people are more devoted to their native foods than the Italians, and Italian groceries filled with imported edibles flourish in all the different colonies of the city. The price of the imported good is a drain on the purses of the patrons and they wearily try to get the same satisfaction out of American-made substitutes, which have the same names and the same appearance, but never, never, the same taste.[7]

To the immigrant palate, Italian-style hams made in America were cured too quickly; American-made *caciocavallo*, the cheese beloved by Sicilians, was lacking in butterfat and quickly spoiled; American garlic was tasteless; and American vinegar was the wrong color. But the saddest disappointment was American pasta, much of it produced on Elizabeth Street in Sicilian-owned pasta factories. Made from standard white flour—not semolina—it was pale and, once cooked, it went soft. Domestic pasta was half the price of imported, but Italians were loath to buy it and literally saved their pennies for the genuine article.

Back home in Italy, peasant families had managed to survive another kind of hostile environment. Oppressive landowners, unfairly high taxes, and periodic crop failures meant a precarious life for the *contadini*, the field workers of southern Italy. Even in good times, when the peasant had enough to eat, starvation remained a looming possibility, always one crop failure away. If nothing else was guaranteed to the peasant—and nothing was—the unshakable bond of family was his bedrock. The family patriarch was an unchallenged authority who demanded absolute obedience from his wife and children, his helpers in the fields. The needs of the family came before those of the individual, and loyalty among family was unwavering. In America, as Italians adapted to a new way of life, the old values of family solidarity were put to the test. For the first time, children left their parents' side to attend school. Here they were exposed to a world of people and ideas apart from their family. Italian girls, no longer under the constant surveillance of their elders, were now free—though not entirely—to make their own friends, and eventually to find their own husbands. As Italian women left their homes to work in American factories, they too developed lives separate from the family, discovering a level of independence they had never known in the Old Country.

Despite all these changes, the old values lived on in the nightly ritual of the evening meal, a tangible expression of family solidarity, loyalty, and love. To borrow a phrase from *Blood of My Blood*, Richard Gambino's wonderful book about Brooklyn Sicilians, the evening meal was "a communion of the family." Sicilians, and southern Italians in general, arrived in America with a deep reverence for the preciousness of food. They knew full well the human labor required to coax it from the earth, and how, on occasion, the earth would refuse them. In America, though now removed from the soil, the Italian still labored for his food, working for relatively low wages in the nation's most strenuous jobs. The family meal was an occasion to share the fruits of that labor, and for the Italian, attendance was mandatory. On weekdays, Italian kids often returned home for lunch as well, though they could eat for free in the school cafeteria.

As Italians found their way into the American economy, the family supper took on another layer of meaning. Edible proof of the immigrant's success, the evening meal was a nightly celebration of the triumph over hunger. The price of that victory was not lost on the immigrants, especially the older ones, who still remembered the fourteen-hour work-days. The bounty before him was the Italian's belated reward for building America's subways, her skyscrapers and bridges—in other words, for bringing America into the twentieth century.

Each night, the family dinner table became a stage for all the tempting foods that the immigrant had once dreamed about but couldn't afford. At the center of that dream, there was meat. For early immigrants, meat was used as a seasoning, an ingredient added to soup or sauce to give it body and richness. By the 1920s, a midweek dinner in a working-class Italian kitchen included soup, then pasta, followed by meat and a salad. At the end of the week, Italian families sat down to a banquet of stunning extravagance. Sunday supper began in the early afternoon with an antipasto of cheese, salami, ham, and anchovies. Appetites now fully awake, the family moved through multiple courses, leading them to the heart of the feast. If the family were Sicilian, that might include a *ragu* made from marrow bones, chicken, pork sausage, and meatballs, stewed veal and peppers, and braciole, a thin filet of pounded beef or pork wrapped around a stuffing of cheese, bread crumbs, parsley, pine nuts, and raisins.

The Italian writer Jerre Mangione, who grew up in Rochester, New York, in the 1920s, remembers the parade of courses: first, there was soup; then pasta, perhaps ziti in *sugo*; followed by two kinds of chicken, one boiled, and one roasted; roasted veal; roasted lamb; and *brusciuluna*—"a combination of Roman cheese, salami, and moon-shaped slivers of hardboiled egg encased in rolls of beef." When the meat was cleared, there was fennel and celery to cleanse the palate, followed by homemade pastries, nuts, fruit, and vermouth. These Sunday suppers, prepared by the author's father, were more extravagant than the family could realistically afford, and the senior Mangione often had to borrow money to pay for them.[8] Moderation had no place at the Sunday table. The gathered crowd, encouraged by their fellow diners, went back for multiple helpings with not a jot of self-consciousness. Quite the contrary, any guest who refused another helping was given a mild rebuke.

The meatballs, *ragus*, and roasts that were (and remain) a centerpiece of the Italian-American kitchen had never figured in the peasant diet. Rather, they belonged to the kitchens of Italian landowners, merchants, and clergy, who, on important feast days like Easter and Christmas, distributed meat to the poor.[9] In America, immigrant cooks reinterpreted these feast-day foods and, in another expression of American bounty, made them a regular part of the Sunday table. The American larder was so immense that it could literally feed the working class on a diet once reserved for Italian nobility.

One effect of the quota laws of the 1920s was to create a shadow wave of uncounted immigrants, men and women from Russia, Poland, Italy, and other restricted nations, who evaded the authorities and entered the country illegally. The Baldizzis belonged to that wave. Adolfo Baldizzi was born in Palermo in 1896 and was orphaned as a young boy. His professional training as a cabinetmaker began at age five, but the outbreak of World War I put his career temporarily on hold. A wartime portrait of Adolfo shows a young soldier with thick black hair and a neatly trimmed mustache. Rosaria Baldizzi (her family name was Mutolo) was born in

Adolfo Baldizzi in his soldier's uniform, circa 1914.

Wedding portrait of sixteen-year-old Rosaria Mutolo Baldizzi, two years before she emigrated.

1906, also in Palermo, to a family of tradespeople and civil servants. The Baldizzis were married in 1922, the same year Rosaria turned sixteen. A year later, the couple decided to emigrate. To evade the 1921 immigrant quota laws, Adolfo came to America as a stowaway aboard a French vessel. As the ship pulled into New York Harbor, he climbed from his hiding place, jumped over the railing, and swam to shore. Rosaria made the same trip in 1924, with a doctored passport.

For Rosaria Mutolo Baldizzi, the move to America was a step down in life. Born to a middle-class family, Rosaria spent her childhood in a good-size house with a stone courtyard where her mother grew geraniums and raised chickens, selling the eggs to neighbors. Rosaria's father was confined to a wheelchair by the time she was a young girl, but in his youth had owned or worked in a bakery. Her two older brothers were both policemen; her sister was a dressmaker. On Sunday afternoons, still in their church clothes, the family took leisurely strolls, stopping at a local café for coffee and granita. Rosaria's marriage to a cabinetmaker at age sixteen was considered a good match. In her 1921 wedding portrait, she is posed at the foot of an ornate staircase, dressed in a tailored skirt and matching jacket, both trimmed in a wide panel of hand-woven lace. A flapper-style cloche hat, jauntily cocked, with a long, trailing sash, completes the ensemble.

Rosaria's move to New York in 1924 meant the end of a reasonably privileged and protected life. Her first glimpse of Elizabeth Street, center of Sicilian New York, was a sobering experience for the young immigrant. To her consternation, the shoppers who overflowed the sidewalk onto the cobblestoned street were oblivious to the garbage under their feet, a carpet of moldering cabbage leaves and orange rinds. Every window ledge and door lintel was veiled in soot, like a dusting of black snow. But most disturbing of all were the Elizabeth Street stables. The young Mrs. Baldizzi was shocked to find that New Yorkers, presumably civilized people, lived side by side with horses.

The couple's first home was a single room in a two-room apartment. To supplement her husband's earnings, Rosaria took in laundry, a common source of income for immigrant women. Whenever the couple fell behind in rent, they simply packed up and moved. The two Baldizzi children, Josephine and John, were born on Elizabeth Street, but in 1928, when John was still in his swaddling clothes, the family left Little Italy for 97 Orchard Street, a leap across cultures that brought the Baldizzis into the heart of the Jewish Lower East Side. Living on Orchard Street, they encountered the challenges typical of an immigrant family. These were eclipsed, however, by the calamitous events of 1929 and their aftermath. The "land of opportunity" they had expected to find evaporated before their eyes, leaving Mr. Baldizzi with a wife, two toddlers, and little hope of finding work.

The Baldizzis remained on Orchard Street through the grimmest years of the Depression. For most of that time, Adolfo was unemployed, though he still earned a few dollars a week as a neighborhood handyman. New clothes or toys for the children were out of the question. When the soles on Josephine's shoes sprouted holes, they were fortified with a cardboard insert. The family food budget was concentrated on a few indispensible staples: bread, pasta, beans, lentils, and olive oil. Once a week, the family received free groceries from Home Relief, the assistance program created by Franklin Roosevelt in 1931 when he was still governor of New York. For many foreign-born Americans, Home Relief introduced the immigrant to foods like oatmeal, butter, American cheese, and, for the children, cod liver oil. It also furnished them with milk, potatoes, vegetables, flour, eggs, meat, and fish. For the Baldizzi parents, the weekly trip to the neighborhood food bank (it was actually the children's school) was a public walk of shame. The food, however, was necessary, and they accepted it gratefully.

Breakfast for the Baldizzi children was hot cereal, courtesy of Home Relief, or day-old bread that Mrs. Baldizzi tore into pieces and soaked in hot milk, with a little butter and sugar. The resulting dish, a kind of breakfast pudding, was a favorite of the children. Josephine Baldizzi, who was always thought too skinny by her parents, was given raw eggs to help fatten her up. The eggs were eaten two ways. Mrs. Baldizzi would poke a hole in one end of the egg, instructing her daughter to suck out the nutritious insides. She also prepared a drink for Josephine made of raw egg and milk whipped together with sugar and a splash of Marsala wine. Breakfast for the parents was hard bread dipped in coffee that Mrs. Baldizzi boiled in a pot. Coffee grinds, like tea bags, were reused two or three times before being consigned to the trash. The Baldizzi children returned home for a lunch of fried eggs and potato, or vegetable frittata. A typical evening meal was pasta and lentils or vegetable soup, which Mr. Baldizzi referred to as "belly wash." On Saturday evenings, he made the family scrambled egg sandwiches with American ketchup.

Though America's bounty eluded the Baldizzis, Rosaria understood the power of food over the human psyche and used it—what little she had—as an antidote to the daily humiliations of poverty. Dinner in the Baldizzi household was a formal event, the table set with the good Italian linen that Rosaria had brought over from Sicily, the napkins ironed and starched so they stood up on their own. If the menu was limited, the food was expertly cooked and regally presented. On occasion, as a treat for the children, Rosaria would arrange their dinners on individual trays and present it to them as edible gifts. One of Josephine's clearest childhood memories is of her mother standing in front of the black stove at 97 Orchard, holding a tray of "pizza"—a large round loaf of Sicilian bread, sliced crosswise like a hamburger bun, rubbed with olive oil, sprinkled with cheese, and baked in the oven. "You see," Rosaria says, "you *are* somebody!" Such was the power of food in the immigrant kitchen: to confer dignity on a skinny tenement kid with cardboard soles in her shoes.

On birthdays and holidays, edible gift–giving rose to the level of ritual. For All Souls Day, when ancestral spirits deigned to visit the living, Mrs. Baldizzi gave the kids a tray piled with the candied almonds known as *confetti, torrone*, Indian nuts, and Josephine's personal favorite, one-cent Hooten Bars. That night, the children would slip the tray under the bed, in case the spirits should arrive hungry. The next day, after the ancestors had presumably helped themselves, it was the children's turn. On Easter, each child received a marzipan representation of the Paschal lamb. Candy was also part of the Nativity scene displayed each December on the Baldizzis' kitchen table, the manger strewn with *confetti* and American-style peppermint drops.

Much of the candy sold in New York in the early twentieth century—Italian and otherwise—was produced locally in factories scattered through the city. In fact, by the turn of the century, New York was the center of the American candy trade, employing more people than any other food-related industry aside from bread. New York candy-makers were tied to another local industry with deep historical roots: the sugar trade. All through the eighteenth and nineteenth centuries, ships carrying raw sugar from the Caribbean delivered their cargo to the refineries or "sugar houses" that were clustered near Wall Street, converting the coarse brown crystals into loaves of white table sugar. Candy factories began to appear in Lower Manhattan in the early 1800s, before migrating uptown to Astor Place, then to midtown and Brooklyn.

By the late nineteenth century, immigrants were important players in the candy business. Some owned factories specializing in their native sweets, but many, many more worked as candy laborers. In New York, Chicago, Boston, and Philadelphia, all major candy-making cities, foreign-born women, chiefly Italians and Poles, worked the assembly lines, dipping, wrapping, and boxing. *Italian Women in Industry*, a study published in 1919 that looked specifically at New York, reported that 94 percent of all Italian working women in 1900 were engaged in some form of manufacturing. While the overwhelming majority worked in the needle trades, about 6 percent were candy workers, a job that required no prior training or skills. The dirtiest, most onerous jobs—peeling coconuts, cracking almonds, and sorting peanuts—went to the older women who spoke no English. For their labor, they were paid roughly $4.50 for a sixty-five-hour workweek. (Women who worked in the needle trades generally earned $8 a week, while sewing machine operators, the most skilled garment workers, brought in $12 weekly or more.) The long hours and filthy conditions in the factories made candy work one of the least desirable jobs for Italian women. Mothers who worked in the candy factories for most of their adult lives prayed that daughters "would never go into it, unless they were forced to."

Another class of candy workers could be found in the tenements. Candy outworkers were immigrant women hired by the factories, who brought their work home to their tenement apartments, completing specific tasks within the larger manufacturing process. This two-tiered arrangement of factory workers and home workers was widespread on the Lower East Side and used by several of the most important local industries. The largest and best-documented was the garment industry, in which factory hands were responsible for the more critical jobs of cutting, sewing, and pressing, while home workers did "finishing work": small, repetitive tasks, like sewing on buttons, stitching buttonholes, and pulling out basting threads, a job that often fell to children. In the candy trade, finishing work meant wrapping candies and boxing them. It also included nut-picking: carefully separating the meat from the nutshell with the help of an improvised tool, like a hairpin or a nail. These were jobs often performed as a family activity, by an Italian mother and her kids, sitting at the kitchen table with a fifty-pound bag of licorice drops lugged home from the factory that morning, a small mountain of boxes at their feet.

Unlike her sisters in the factories, the home worker fell beyond the reach of protective labor laws that regulated the length of her workday along with her minimum wage. Her children were also unprotected. In the Old Country, children worked side by side with their fathers in the fields. In cities like New York, the moment they returned from school they went to work shelling hazelnuts or walnuts until deep into the night. During the rush seasons just before Christmas and Easter, they were kept home from school entirely.

In the eyes of middle-class America, the candy home worker was an ambiguous figure, equal parts victim and villain. Bullied and abused by greedy factory owners, she attracted support from social reformers like the National Consumer League, which advocated on her behalf. At the same time, the outworker was a threat to public safety, the foods that touched her hands contaminated by the same germs that flourished in her tenement home. Pasta, wine, matzoh, and pickles were also produced in the tenements, foods made by immigrants for immigrants. Candy, however, was different. While made by foreigners, it was destined for the wider public, available in the most exclusive uptown stores.

As middle-class Americans became aware of tenement candy work-

ers, panic set in. Tenement-made candy, they surmised, was the perfect
vehicle for transporting working-class diseases like cholera and tuberculosis from the downtown slums to the more pristine neighborhoods
in Upper Manhattan. "Table Tidbits Prepared Under Revolting Conditions," a story that ran in the *New York Tribune* in 1913, sounded the alarm:

> *Foodstuffs prepared in tenement houses? For whom? For you, fastidi-*
> *ous reader and for everybody! A pleasant subject this for meditation.*
> *Slum squalor has been reaching uptown in many insidious ways. It*
> *was bad enough to think that the clothing one wore had been handled*
> *in stuffy rooms, where sanitary conditions and ventilation were de-*
> *plorable . . . When it is learned, however, that many of the things*
> *actually eaten or put to the lips have been prepared by some poor*
> *slattern in indifferent or bad health and by more or less dirty tots of*
> *the slums amid surroundings that would cause humanity to hold its*
> *nose, a brilliant future looms up for some of the scourges scientists are*
> *busily endeavoring to stamp out.*[10]

An immigrant family shelling nuts at their kitchen table. This photo was used to demonstrate the
unsanitary conditions that prevailed among tenement home workers.

Stories like this one, brimming over in lurid details, enumerated the many sanitary breaches committed by the home worker. She was known, for example, to crack nuts with her teeth and pick them with her fingernails. At mealtime, picked nuts were swept to the side of the table, or removed to the floor or the bed. But most alarming of all, home workers performed their tasks in rooms shared by tuberculosis sufferers or children sick with measles. In some cases, the home worker was sick herself, like the consumptive woman who was too weak to leave her bed but somehow managed to go on with her work as a cigarette maker. The one detail consistently absent from these stories was mention of any documented illness linked to tenement-made goods. They may have occurred, but the looming threat posed by the home worker was more compelling than the actual risk.

While immigrant candy workers fed the national sweet tooth, in their own communities, Italian confectioners made sweets for their fellow countrymen. The more prosperous owned their own shops, or *bottege di confetti*, preparing the candy in large copper pots at the back of the store. Marzipan, *torrone* (a nougat-like candy made with egg whites and honey), and *panforte*, a dense cake made of honey, nuts, and fruit, were specialties of the confectioner's art. The *confetti* shops also sold pastries like cannoli and *cassata*, an ornate Sicilian cake made with ricotta, candied fruit, sponge cake, and marzipan. In the months leading up to Easter, store owners created window displays of their gaudiest, most eye-catching sweets. Herds of marzipan lambs grazed in one corner of the window, beside a field of cannoli. Pyramids of marzipan fruits and vegetables, each crafted in fine detail, loomed in the background. Most eye-catching of all, however, were candy statues representing the main actors in the Passion story: a weeping Virgin Mary in her blue cloak; Christ in his loincloth staggering under the weight of the cross; Mary Magdalene; and even the heartless Roman soldiers brandishing their spears, all cast from molten sugar.

The Easter celebration was a family event centered on the home. More conspicuous occasions for candy consumption were the religious festivals held through the year in the streets of Little Italy. The *feste* were

Nut peddler at an Italian street festival.

open-air celebrations in honor of a particular saint, each one connected with a town or village back home. Sponsored by fellow townspeople, they combined religious observance in the form of a solemn procession with brass bands, fireworks, and public feasting, the exact nature of food determined by the immigrants' birthplace. A Sicilian festival, for example, would include *torrone*, but not the kind made with egg whites. Sicilian *torrone* was a glossy nut brittle made with almonds or hazelnuts, a confection brought to Sicily by the Arabs. There was also *cubbaita*, or sesame brittle, another Arab sweet, and *insolde*, a Sicilian version of *panforte*. Below is a description of the foods, candy included, available at a 1903 festival held in Harlem, the uptown Little Italy, honoring Our Lady of Mount Carmel:

> *The crowd is chiefly buying things to eat from street vendors. Men push through the masses of people on the sidewalks, carrying trays full of brick ice cream of brilliant hues and yelling "Gelati Italiani"—Italian ices. "Lupini," the "ginney beans" of the New York Arab; "cice-retti," the little roasted peas and squash seeds are favorite refreshments.*

Great ropes of Brazil nuts soaked in water and threaded on a string, or roasted chestnuts, strung in the same way, lie around the vendor's neck. Boys carry long sticks strung with rings of bread. All manner of "biscuitini," small Italian cakes, are for sale, frosted in gorgeous hues, chiefly a bright magenta cheerful to look upon but rather ghastly to contemplate as an article of food.

Boys at the door of bakeshops vociferate "Pizzarelli caldi"—hot pizzarelli. The pizzarello is a little flat cake of fried dough, probably the Neapolitan equivalent of a doughnut. They sell for a penny a piece. Sometimes the cook makes them as big as the frying pan, putting in tomato and cheese—a mixture beloved of all Italians. These big ones cost 15 cents, but there is enough for a taste all around the family. The bakers are frying them hot all through the feast. A certain cake made with molasses, and full of peanuts or almonds, baked in a long slab and cut in little squares, four or five for a cent, is much eaten. So is "coppetta," a thick, hard white candy full of nuts; and the children all carry bags of "confetti," little bright-colored candies with nuts inside. Here and there the sun flashes on great bunches of bright, new tin pails, heaped on the back and shoulders of the vendor: and the new pail bought and filled with lemonade passes impartially from lip to lip of the family parties lunching on the benches in Thomas Jefferson Park.[11]

Below is a recipe for *croccante*, or almond brittle. It is adapted from *The Italian Cook Book* by Maria Gentile, published in 1919, among the earliest Italian cookbooks published in the United States.

CROCCANTE

3 cups blanched sliced almonds
2 cups sugar
1–2 tablespoons mildly flavored vegetable oil
1 lemon cut in half

Preheat to 400°F. Liberally grease a baking tray with the
vegetable oil and put aside. Spread almonds on a separate
tray and toast in hot oven until golden, about 5 minutes.
Heat sugar in a heavy-bottomed saucepan and cook until
sugar has completely melted. Add almonds and stir. Pour
hot mixture onto greased baking tray, using the cut side of
the lemon to spread evenly. Allow to cool and break into
pieces.[12]

In the descriptive names that immigrants invented for the United States
is a measure of what they expected to find. To Eastern European Jews,
America was the *Goldene Medina*, or "Golden Land," a place of extravagant wealth. For the Chinese immigrants who settled on the West Coast,
America was "the gold mountain," a reference to the California hills that
would make them rich. Sicilians, by contrast, referred to America as "the
land of bread and work," an image of grim survival, comparatively speaking. To the Sicilian, however, bread was its own form of wealth. More than
other Italians, Sicilians felt a special closeness to this elemental food, a
"God-bequeathed friend," in the words of Jerre Mangione, "who would
keep bodies and souls together when nothing else would."[13] The Sicilian
respect for bread was rooted in a long history. From the sixteenth century
forward, bread formed the axis of the peasant diet, sustaining—though
just barely—generations of Sicilian field workers. The typical Sicilian
loaf was made from a locally cultivated strain of wheat, *triticum durum*,
which Arab settlers had brought to Sicily, along with almonds, lemons,
oranges, and sugar, back in the ninth century. The wheat grew on giant
estates called *latifundi*, and by the Middle Ages it covered most of Sicily's
arable land. Durum was a particularly hardy strain with a high protein
content, producing a dense, chewy bread with a powerful crust. When
mixed with water, it could be stretched into thin sheets that resisted tear-
ing, making it ideal for pasta, too.

Peasants who worked all day in the field packed a hunk of bread and maybe an onion for their lunch. For dinner, there was bean soup, and yet more bread. As minimal as this sounds, the Sicilian pantry became even more spare during periods of famine, which in Sicily amounted to "a time without bread." A Sicilian proverb recounted by Mary Taylor Simeti in *Pomp and Sustenance* sums it up beautifully:

> *If I had a saucepan, water, and salt,*
> *I'd make a bread stew——if I had bread.*

A loaf of bread for Sicilians embodied the basic goodness of life. Where we might say a person is "as good as gold," a Sicilian says "as good as bread." A piece of bread that fell to the ground was kissed, like a child with a scraped knee.

When Sicilians described America as the land of bread and work, they imagined a country without hunger, which, in their experience, was just as miraculous as a city paved in gold. One thing they never imagined, however, was that bread would fall into their hands like manna from heaven. Richard Gambino, describing the powerful work ethic that guided his Brooklyn neighborhood, remembers a phrase often repeated by the local men, *"In questa vita si fa uva,"* literally translated as, "In this life, one produces grapes." In other words, each one of us has been put on this earth to be useful. Gambino also remembers his grandfather's Sicilian friends holding up their calloused workers' hands and saying *"America e icca,"* meaning "America is here—this is America."[14] For the Sicilian, bread and work were locked together in a kind of dance that began early in life, the day the young Sicilian was old enough to contribute.

Sicilians carried their bread tradition to America, where it continued with certain necessary revisions. On Saturdays in Sicily, women did all their baking for the coming week, a way to preserve precious fuel. In cities like New York, the Italian laborer, now on his own, purchased his weekly ration each Saturday in Little Italy. Here, Italian peddlers sold loaves prepared expressly for the "diggers and ditchers," immense crusty loaves nearly the size of a wagon wheel. Anyone who could not afford to buy their bread fresh, bought stale loaves, a special sideline of the retail

Curbside bread peddler on Mulberry Street, selling her wares from a basket.

bread trade controlled by women. Bread that was once thrown away was sold by the larger bakeries and retail houses to middlemen who, in turn, sold it to the peddlers, women who sat on the curbstone, their goods piled in a blanket beside them. As the immigrant's finances swelled, his diet branched out in new directions. Pasta, eggs, and, eventually, meat diverted some of his affection for bread but never replaced it. In the Mangione household, bread was eaten with every course of the Sunday feast, except for the pasta. A bowl of soup without bread was bereft of its faithful companion. Meat without bread was considered sinful.

Elizabeth Street in the 1930s hosted two groceries, two butcher shops, one fish store, and one candy store, but six bakeries. In New York, the typical Sicilian loaf was a simple round bread sprinkled with sesame seeds, weighing roughly two pounds. On holidays and feast days, however, the baker's imagination took flight, and the Elizabeth shops offered fantastic bread sculptures, each shape tied to a particular saint. One resembled a curving bowl of flowers, another was shaped like a swirling backward *S* with frilled edges. Though each of these fanciful shapes was created for the saints, the baker hedged his bets by including breads that were braided or knotted, traditional forms of protection from the evil eye.

When they lived on Orchard Street, the Baldizzis relied on bread as a food of survival in much the same way it had been for generations of Sicilian peasants. It was eaten at every meal, and for breakfast it *was* the meal. Stale or hard bread was never thrown away. Instead, Rosaria would rub it with a little water and oil and put it in a warm oven to soften it up. Bread that was too far gone to revive was turned into bread crumbs, the indispensable ally of all Sicilian cooks. Bread crumbs were used to stretch more costly ingredients like meat, and sometimes to replace it entirely. Combined with oil, parsley, and garlic, bread crumbs were used as a stuffing for peppers, zucchini, artichokes, and other vegetables. Bread crumbs, parsley, and eggs were used to make frittata, a standard mid-day meal for the Italian laborers who dug the New York subway system. Bread crumbs were also toasted in hot oil and sprinkled over pasta or pizza as a replacement for the more expensive grated cheese. Even as their incomes rose, Sicilian-Americans continued to cook with bread crumbs, a former food of necessity. Nothing made a crunchier coating for fried calamari or *arancini*, the creamy, fist-size rice balls filled with ground meat or cheese, sold as street food and prepared in the home for holidays and parties. Today, they appear on the menus of old-time Sicilian restaurants and delicatessens. Below are two bread crumb recipes. The first is for a bread-crumb frittata.

ZUCCHINI FRITTATA

4 large or 6 small zucchini
1 small onion sliced
3 tablespoons olive oil
5 eggs
½ cup grated Romano cheese
⅓ cup bread crumbs
Salt and pepper

Rinse and grate zucchini using the large holes on a hand-held grater. You could also use the shredding disc on a food processor. Place grated zucchini in a colander over the sink. Sprinkle with a teaspoon of salt, toss, and let sit 15 minutes or longer, until the zucchini begins to "weep." Squeeze out the extra moisture with your hands. Put aside.

Sauté the onion in 2 tablespoons of the olive oil until lightly golden. Add the zucchini. Cook until zucchini is slightly browned in spots and most of the remaining moisture has cooked away. Season with salt and pepper, and remove from the stove. Meanwhile, beat the eggs in a large mixing bowl. Add the warm zucchini, along with the grated cheese and bread crumbs. Let the mixture sit a minute or two, so the bread crumbs can drink up some of the egg. Add remaining oil to the frying pan, and let it get hot. Add egg mixture, then turn down the heat. Cook over a low flame until frittata starts to set around the edges. Place under a hot broiler to finish cooking the top. Slide onto a plate. Eat at room temperature.[15]

The pasta recipe below comes to us from Concetta Rizzolo, an immigrant from Avellino, a town east of Naples, who settled in New Jersey in the 1910s. The toasted bread crumbs can be stored in the refrigerator for several weeks.

SPAGHETTI CON AGLIO E OLIO

½ cup olive oil
6 cloves garlic, minced
½ to 1 tsp red pepper flakes
10 black peppercorns
1 pound spaghetti

½ cup chopped parsley
½ cup toasted bread crumbs

To toast bread crumbs, coat a frying pan with olive oil and
heat over a medium flame. When the oil is warm, add I
cup homemade bread crumbs. Stir bread crumbs to prevent
them from burning or sticking. They are ready as soon as
they turn a uniform golden-brown.

Boil water for spaghetti. Meanwhile, heat oil over low
flame. Add garlic and peppercorns. Cook garlic over a very
low flame until it is soft and translucent, but do not allow
it to brown. Add pepper flakes and continue to cook for
about five minutes.

Drain pasta but reserve about a cup of cooking water.
Add to oil and garlic mixture and heat over low flame. Toss
with spaghetti and serve immediately with parsley and
toasted bread crumbs.[16]

Living on Orchard Street, in the heart of the Jewish East Side, the
Baldizzis formed close relationships with their Jewish neighbors, Fannie Rogarshevsky in particular. After her husband died, when Fannie
assumed the role of building janitor, Mr. Baldizzi, a trained carpenter,
often helped her with repairs. The two became good friends. Sometime
in the early 1930s, to alleviate crowding in the Rogarshevsky household,
one of Fannie's grandsons was sent to live with the Baldizzis. Conversely,
Fannie regularly fixed school-day lunches for the Baldizzi kids when their
mother was at work. (Rosaria Baldizzi worked for a short time during the
Depression but had to give up her job or forfeit her check from Home
Relief.) The young Josephine, Fannie Rogarshevsky's designated *shabbos
goy*, was fascinated by the way the Jewish homemaker koshered her chickens, scrubbing them at the kitchen sink the way an Italian mother might
scrub an especially dirty child. Every evening, Mr. Baldizzi stopped by

Schreiber's Delicatessen on Broome Street for a glass of schnapps, his ritual nightcap. On weekends he took the kids on long walks that carried them over the Manhattan Bridge and back again, stopping along the way for hot potato pancakes, that quintessentially Jewish snack sold by the East Side vendors. Even so, when it was time to shop for groceries, Mrs. Baldizzi found limited use for the pushcart market directly below her window. Instead, the food she depended on could be found a few blocks away, in the Italian pushcart market on Mulberry Street. By the time of the mayor's pushcart commission in 1906, Mulberry Street was already a full-service open-air market catering to the Italian homemaker. Satellite markets sprang up on Elizabeth and Bleecker streets and along First Avenue below 14th Street. The largest of all, however, extended for nearly a mile from 100th to 119th streets on First Avenue in East Harlem. The pushcart markets were the single most important source

Clams on the half-shell, a common street food in Little Italy.

of food for immigrant New Yorkers from the 1880s through the late 1930s, when Mayor La Guardia, the son of immigrants himself, finally prevailed in the decades-long battle between the pushcart peddlers and the city government. In anticipation of the upcoming World's Fair and the multitudes that would soon descend on New York, La Guardia shut down one market after another, consolidating some and moving others into newly built market buildings more befitting a modern city. The Mulberry Street market fell to the mayor's ax in 1939, the year the fair opened. In the meantime, however, the Mulberry Street pushcarts supplied Italian cooks like Mrs. Baldizzi with foods unknown in the Jewish quarter. The Italian peddlers, for example, did a brisk business in mussels, periwinkles, conch, oysters, and clams. The last two items were sold as street food, the Italian equivalent to the Jewish knish.

In the Old Country, Italian women foraged for snails, a free source of precious protein. In America, the snail peddler became a fixture of the Italian pushcart market. His mode of advertising, unique among street vendors, was an upright board with the snails clinging to it. The bulk of his delicate stock could be found in a crate under the pushcart to keep the snails shaded and cool. Italian cooks soaked their snails in cold water, prompting the animal to inch out of its shell, then fried it with garlic, creating a savory and inexpensive topping for pasta. Other sea creatures sold at the market were squid, octopus, and eels, which the Italian cook stewed with tomatoes.

But the most remarkable feature of the Italian markets was the selection of greens and other vegetables that figured so prominently in the immigrants' diet. Early accounts of the pushcart markets offer a partial list of the many forms of plant life sold by Italian peddlers. There was cabbage and cauliflower—though no mention of broccoli—cow peas, cucumbers, celery (distinct from American celery), fennel, peppers, tomatoes, eggplant, onion, garlic, chickweed, beet tops, lettuce, and many other forms of leafy greens which the Americans had no names for. As a convenience to the homemaker, beans could be purchased dried or already soaked to hasten the cooking time. The Italian vegetable peddlers, many of them women, were fastidious in the presentation of their goods.

As one observer noted, "They clean, cut, and freshen the vegetables, constantly rearranging them so that they appear to the best advantage on the stands."[17] The women scrubbed their celery and their fennel until the stalks gleamed; they buffed the peppers and sprinkled the lettuce with water to keep it from wilting. Americans, who still thought of salads as an assemblage of vegetables and meats, often bound with mayonnaise, were struck by the Italians' more restrained approach to the same dish. "No other people in the world," one observer remarked,

> *use so many salads. They never mix tomatoes, lettuce, and cucumbers as Americans do in their salads. Each is kept separate. For their tomato salad, they clean the tomatoes, but do not peel them, split or slice them, and dress them with oil, but no vinegar. Then they strew over them stems of "regona," a herb of aromatic taste and smell which comes dried from Italy . . . Cucumbers they peel and eat with both oil and vinegar, regona, and garlic. The heart of the lettuce, which they call "lattuga," is a prime favorite, dressed with olive oil and Italian vinegar.*[18]

The Italian's devotion to these simple preparations was also noteworthy. "The Italian invariably has a salad for dinner if he can afford it," our observer adds, "and it seems often to supply the place of meat to him." To the American, the Italian salad habit was a source of puzzlement. How could a plate of lettuce, they wondered, take the place of a good roast? More curious still, how could a diet so lacking in substance provide the nourishment required to sustain human life?

Come spring, immigrants scoured the vacant lots of Brooklyn and the Bronx for wild dandelions, a food they had once gathered in the fields around their native villages. In New York, dandelions were a source of income for Italian women who collected the greens, cleaned them, and sold them at the pushcart markets, a washtub's worth for a nickel. Italian cooks used the wild green for dandelion soup, or fried them in olive oil with tomatoes and pepper. When boiled for several hours, the filtered cooking liquid, *"acqua di cicorie"* was used as a tonic for "dyspepsia and general weakness."[19] An Italian summer delicacy was *cucuzza*, an extremely long squash with smooth, pale green skin and a hook at one end like an umbrella handle. Peddlers displayed the *cucuzze* by looping the hooked ends over a horizontal pole so they hung like stockings on a wash line. The leaves of the plant, sold as a separate vegetable, were heaped in crates on the pushcart below. *Cucuzza* was especially popular with Sicilian cooks, who added it to soups or fried it with garlic and tomatoes. Sicilian confectioners, meanwhile, chopped the squash into small pieces or shaved it into ribbons, then boiled it in sugar syrup until the opaque white interior had turned deep gold and was almost transparent.

Much of the produce sold on Mulberry Street was grown on immigrant "truck farms" in Brooklyn, Queens, Long Island, and New Jersey, a few hours commute from the wholesale markets in Lower Manhattan. A bit farther out from the city, Italians established a vibrant farming community in Vineland, New Jersey, which by 1900 comprised two hundred and sixty immigrant families. The Vineland farmers cultivated fruits and vegetables for the immigrant market using seeds imported from Italy. Their crops included garlic, peppers, cauliflower, cabbage, beets, fennel, cardoons (a relative of artichokes), chestnuts, figs, plums, and ten varieties of grapes.

In the nineteenth century, immigrant Jews carried their goose-farming tradition to America, establishing urban poultry farms in the East Side tenements. Several decades later, Italian immigrants brought their treasured home gardens to urban America, now reconfigured as a tenement window box. In wooden planters made from discarded soapboxes, Italian homemakers grew oregano, basil, mint, peppers, tomatoes, and lettuce. (The more ambitious urban farmers planted their gardens on tenement rooftops.) The tradition of the home garden continues today in Italian neighborhoods like Hoboken, New Jersey, and Bensonhurst, Brooklyn, where second- and third-generation immigrants grow basil and plum tomatoes in emptied institutional-size cans of *pomodori pelati*.

Though removed from the soil, transplanted Italian women moved to the rhythm of the agricultural year. Each fall, in New York's Little Italy, they bought up great loads of peppers and preserved them for the coming winter. The peppers were split and brined in tubs of saltwater or packed in jars filled with vinegar. The women also dried their peppers in the sun, just as they used to in the Old Country. In New York, however, they threaded the peppers onto long strings and suspended them from the fire escapes in great dangling loops. Tomatoes were also dried in the sun, along with eggplant, which the women first cut into strips. Each of these dried vegetables was soaked in water for several hours prior to cooking, then fried in olive oil or added to soup. The eggplant recipe below is from Maria Gentile.

EGGPLANTS IN THE OVEN

Skin five or six eggplants, cut them in round slices, and
salt them so that they throw out the water that they contain.
After a few hours, dip in flour and frying oil.

Take a fireproof vase or baking tin and place the slices
in layers, with grated cheese between each layer, abundantly
seasoned with tomato sauce.

Beat one egg with a pinch of salt, a tablespoonful of
tomato sauce, a teaspoonful of grated cheese and two of
crumbs of bread, and cover the upper layer with this sauce.
Put the vase in the oven and when the egg is coagulated,
serve hot.[20]

The contempt for Italian cooking that prevailed in this country a hundred-plus years ago is a buried fact in our culinary history and a surprising one, too, considering how much attitudes have changed. In the United States today, no immigrant cuisine is more embraced by the American cook, her kitchen stocked with tomato paste, canned tomatoes, jarred marinara sauce, olive oil, parmesan cheese, garlic, and above all pasta, mainstay of the American dinner table. And what food, if not pizza, is more beloved by American schoolchildren?

This national love affair unfolded in two overlapping yet discon-
nected chapters. Chapter one began in the mid-nineteenth century, as
immigrants from northern Italy settled in New York, Philadelphia, Baltimore, New Orleans, Boston, and San Francisco. Italians who belonged to
this first wave were largely people of culture—artists, musicians, teachers, doctors, and other professionals. Among the early immigrants were
restaurant-keepers. In the 1850s and 1860s, as the Italian settlements
gathered critical mass, they opened eating places to feed their transplanted countrymen. New York's first Italian restaurants were clustered
near Union Square, close to 14th Street, then the city's main entertainment thoroughfare and home to the Academy of Music, the nation's first
official opera house. In 1857, an Italian named Stefano Moretti opened
a pleasingly shabby second-floor restaurant directly across the street from
the Academy, which became New York's first important Bohemian dining
spot. The favorite haunt of Italian opera stars, Moretti's began to attract
native-born musicians, artists, and writers. The avant-garde of their day,
they were drawn to Moretti's by the delightfully foreign atmosphere as
well as the food, a five-course dinner for a dollar, a fair price for the time
though well beyond the means of the working class. The dishes they
encountered were typical of the northern kitchen. Risotto with kidney
was a house specialty, along with wild duck and quail, both served with
salad. But Signor Moretti was also known for his spaghetti, "tender as
first love" and "sweet beyond comparison." Though nineteenth-century
Americans were generally familiar with macaroni, the pipe-stem-shaped
pasta used today for mac 'n' cheese, spaghetti was still utterly alien. American diners were simultaneously baffled, alarmed, and enchanted by these
attenuated strands of dough that they discovered in Italian restaurants
but still had no name for. If the food itself was bizarre, the complicated
procedure of eating it left Americans awestruck. The following description of a New York Italian restaurant circa 1889 captures the air of
adventure surrounding this novel food. The second course consisted

> of a substance resembling macaroni that has been pulled out until each
> piece is at least two feet long, while the thickness has proportionally di-

minished. You are told that it is wholly bad form to cut this reptile-like
food; you must eat as the Italians do. Thereupon you suddenly cease
to feel hungry, and spend the time in observation. They, to the manner
born, lift a mass of this slippery thing upon the fork, give the wrist
several expert twists, and then, with lightning rapidity, place it in the
mouth. If by misfortune a string escapes, it is gradually recovered in
the most nonchalant manner imaginable. It is a fascinating operation,
though by no means one to inspire a desire to emulate the operator.[21]

The only thing more diverting than this queer new food was the foreign crowd, a collection of singers, ballerinas, professors, journalists, and
businessmen.

As the city's theaters migrated from 14th Street to Broadway, the restaurants followed. For New Yorkers out on the town, tired of their native
chop houses and oyster saloons, Italian food was a refreshing change of
pace, and much cheaper than French, the foreign cuisine favored by elite
society. Italian food, by contrast, along with German, was the foreign
cuisine of the American middle class. The kind of food New Yorkers
could expect to find at the new Broadway restaurants was more attuned
to American tastes. For the less adventuresome eater, they offered both
chops and oysters. The rest of the menu was still rooted in the more
mildly flavored and buttery cuisine of northern Italy. An inventory of
recommended dishes that ran in the *New York Sun* advised diners to stick
with the cheaper items on the menu, like spaghetti in meat sauce, "a
chopped-up soupy compound," and to skip the more expensive meats in
favor of veal, lamb, and giblets:

The leg of veal, usually used for a soup bone, is delicious when it
comes on the table as osso buco. The leg is roasted, there is a suggestion
of herbs and garlic and a sauce of brown butter over the risotto that
accompanies it. Then there is the delicious marrow in the bone that has
been opened in order that it may easily be eaten.

The veal cutlets, whether they are served à la Milanaise, with cêpes
cut up over them and put into a sauce of butter and cheese, or with

herbs, are superior to any that can be eaten at the best of Fifth avenue restaurants.

Arostino, which means a little roast, is a slice of the veal served with the kidney embedded into it and cooked with thyme and a thick brown sauce covering it and a bed of risotto. Such a cut of veal is unknown to American butchers.

The kidneys au sauce Medere are made in accordance with an Italian formula and are remarkable from the fact that only very small kidneys are used and they are served with champignons of about the same size.

It is in such dishes as these that the Italian restaurants excel, and to them they owe their present popularity, for they alone are able to serve them in such excellence in cooking and at such prices.[22]

The Broadway restaurants offered just enough novelty (and garlic) to stimulate the imagination while providing diners with the niceties of New York's finest eating establishments, including an Italian menu printed in French.

After 1880, an entirely different kind of Italian eating place could be found in New York. These were the basement restaurants on Mulberry and Mott streets that catered to the new wave of Italian immigrants, peasant farmers from the southern half of the country who began to settle in New York's notorious Five Points neighborhood, moving into the ramshackle tenements once occupied by American blacks and poor Irish. The Italians also took over the low-paying jobs once held by the Irish to become the city's new street cleaners and ditch diggers. Native-born New Yorkers drew a firm distinction between these new immigrants and the Italians they already knew, "honest," "industrious," and "orderly" people. An 1875 editorial that ran in the *New York Times* presented this thumbnail portrait of the new arrivals:

They are extremely ignorant, and have been reared in the belief that brigandage is a manly occupation, and that assassination is the natural sequence of the most trivial quarrel. They are miserably poor, and it is not strange that they resort to theft and robbery. It is, perhaps, hopeless to think of civilizing them.[23]

While the north vs. south distinction was rooted in historical fact (southerners *were* poor and uneducated), it became the foundation for pernicious stereotypes imposed on southern Italians, which Americans returned to again and again, using the immigrants' birthplace to explain everything about them, from their violent nature to their deplorable eating habits. As a result, several years passed before Americans were able to gather up their courage and sample the fruits of the southerners' kitchen.

The first restaurants in Little Italy reflected the immigrants' meager earnings. A visitor to the Italian colony in 1884 counted four neighborhood restaurants where laborers could buy a two-cent plate of macaroni, three cents worth of coffee and bread, or splurge on coffee and mutton chops, the most expensive item on the menu, for a total of six cents. Basement restaurants also provided laborers with a place to gather, to smoke their pipes, play cards, and enjoy the talents of their musical peers. (The neighborhood's busiest social spots, however, were the stale-beer dives, as they were known, where the house drink was made from the beer dregs collected from a better class of saloons.) The dining rooms attached to boardinghouses and cheap hotels were another eating option for the transplanted Italian. A typical menu consisted of coffee with anisette and hard bread for breakfast, and for supper, minestrone, spaghetti or macaroni, followed by a stew made with garlic and oil.

The early restaurants reflected the strong connections that immigrants felt for their native villages. As they settled in New York, Italians recreated the geography of home, with Neapolitans on Mulberry Street, Calabrians on Mott, Sicilians on Elizabeth, and so on. Within these regional encampments, Italians from a particular town or village tended to cluster on the same city block and sometimes in the same building. At their *festa*, villagers came together to honor their local saint, but also to celebrate their ties with each other. Italians have a word for the special connectedness felt among townspeople. *Campanilismo*, from the Italian word for "bell," describes the bonds of solidarity felt among people who live within hearing distance of the same church bell.

Restaurants preserved these regional loyalties. Some of the first restaurants were hidden within the Italian groceries that began to appear in New York in the 1880s, the provisions lined up on one half of the room, the other half set aside for tables. Like many immigrant restaurants, these were family-run businesses. The store/café occupied the front room of a ground-floor apartment, while the family slept in the back. This was also where the proprietor's wife cooked for her customers. An 1889 article from *Harper's* magazine describes the convivial scene inside one of these store/cafés, this one owned by a family of Sicilians:

> *Notwithstanding the poverty of the place, it is as busy as a beehive. At the long table, a number of men, who probably work at night on the scows of the Street-cleaning Department, are drinking the black coffee, which, despite its cheapness, is palatable enough to the drinkers. A handful of Italian women, whose dresses and shawls are bright with the gaudy colors so dear to them, are chaffering with the proprietor's wife over a string of garlic or a pound of sausage. The chairs about the room are occupied by friends and customers of the house, who are smoking villainous short pipes and talking so loudly that one ignorant of the language would suspect them to be on the point of a riot.*
>
> *The air is blue with tobacco smoke, and the place reeks with the conglomeration of stenches that no language can describe, yet all the people appear to enjoy the best of health, and even the children display a robustness and physical vigor that would do credit to those born with silver spoons. The food served in this, as in all places of a similar sort, does not lack nutrition, though the materials gathered would not recommend themselves to the fastidious. The stew, made up of scraps gathered here and there, is spiced until savory to a hungry man, and the macaroni, though manufactured from the cheapest and coarsest flour in some eastside shop, is usually wholesome.*[24]

If Americans were charmed by the Italians' earthiness, an establishment like the Sicilian café was best experienced in the pages of magazines.

Once confined to the Five Points, New York's most notorious slum, by 1910 Little Italy stretched north toward Houston Street and west toward Greenwich Village. As the colony expanded, its physical character also changed. In 1895, the tenements surrounding Mulberry Bend, the heart of the Five Points, were torn down to make way for Columbus Park, Little Italy's new "town square." Two blocks west of the Bend, more tenements were razed to widen Elm Street, creating a broad thoroughfare known today as Lafayette Street. Now open to light and air, Little Italy was no longer the "foul core of New York slums" that Jacob Riis described in the 1890s. What's more, Little Italy was no longer the bachelor community it once had been. By 1900, Italian immigrants were largely men and women who came to the United States to start a family and lay down roots.

Chapter two of the American romance with Italian food began at the turn of the century, as native New Yorkers wandered into the now-expanded Italian colony. One stop on their itinerary was the Italian grocery, which exposed the visitor to enticements they had never known existed. The heart of their education, however, took place in the Italian restaurants that served as culinary classrooms. For Americans who believed that Italians subsisted on bread and macaroni, the edible delights available in the downtown restaurants came as a revelation. Published accounts of these gastronomic forays were quick to warn readers of certain possible pitfalls. To quote one newspaper reporter: "The Italian taste in cookery is not always such as pleases the native American palate."[25] Properly advised, however, American diners could find an array of delectable dishes: minestrone that was "thick and tasteful"; kidneys, liver, and veal, prepared southern-style with peppers and onions; simmered *polpette*, or meatballs; fried calamari; and a host of tantalizing vegetable dishes based on eggplant, tomatoes, and peppers. But the dish that most enchanted American diners was spaghetti.

Discovered by an earlier generation of adventuresome eaters, in the early 1900s spaghetti reentered the American culinary consciousness and quickly moved from restaurant kitchens to the family dinner table. Recipes for spaghetti began to appear in American cookbooks, including *The Boston Cooking School Cook Book* by Fannie Farmer published in 1896. In her 1902 cookbook, Sarah Tyson Rorer, a leading voice of the domestic science movement, explains precisely how this still-novel food should be cooked:

> *Spaghetti is always served in the long form in which it is purchased. Grasp the given quantity in your hand; put the ends down into boiling water; as they soften, press gently until the whole length is in the water; boil rapidly for twenty minutes. Drain, and blanch in cold water.*[26]

Around the same time, recipes for spaghetti began to appear on the women's pages of American newspapers. Many sent in by readers, newspaper recipes leave a vivid record of the creative, often zany applications that American home cooks found for spaghetti. There was "Mexican Spaghetti" with tomatoes, paprika, peppers, and bacon served in a chafing dish; "Chicken and Spaghetti Croquettes" made with cooked spaghetti, finely chopped; and the popular "Tomatoes Stuffed with Spaghetti." In 1908, the women's page of the *Chicago Tribune* featured a reader's recipe for "Spaghetti and Meat Balls," one of the earliest references to this future staple of the Italian-American kitchen:

SPAGHETTI AND MEAT BALLS

Take one pound of round steak, run through meat grinder two or three times; one egg, three rolled crackers or grated stale bread, one small onion grated, four sprigs parsley chopped fine, and pepper and salt to suit taste. Mix and form into small balls, a teaspoonful and a half each.

Prepare sauce as follows: One can tomatoes, one green or red pepper, one onion, two bay leaves, and a quart water. Boil one hour, then strain through colander. Add small piece of butter, and pepper and salt to suit taste. Return to fire and place meatballs in it and boil slowly for forty minutes.

Spaghetti: take one pound of spaghetti, boil it in two quarts of saltwater for twenty minutes, drain, pour over sauce and all, and serve hot.[27]

The Baldizzis' financial prospects improved considerably when America entered the war in 1941. By this time, the family was living in Brooklyn. Adolfo found work in the wartime naval yards, while Rosaria returned to her job in the garment district. With both parents employed full-time, the pall cast over the Baldizzi household began to lift. On Orchard Street, the family had owned a radio, which Rosaria kept tuned to the opera stations.

Josephine and John Baldizzi on the roof of 97 Orchard Street, 1935.

In Brooklyn, she bought a record player and kept it running whenever she was home, filling the house with music. On holidays, the Baldizzis hosted family parties complete with music and dancing. The new prosperity brought very welcome changes to the family dinner table. At last, Rosaria could afford to buy the meat that was so conspicuously absent from the Orchard Street kitchen. The most festive meal of the year, however, was entirely meatless. Christmas dinner traditionally began at ten o'clock, to coordinate with the midnight mass. The first course was octopus salad, followed by a pan of lasagna, rich with ricotta, eggs, and mozzarella, but the climax of the meal was a stew made with *baccala*—salt cod. Here is the Baldizzi family recipe for the holiday staple:

CHRISTMAS *BACCALA*

1 stalk celery, diced
½ cup chopped onion
2 cloves garlic, chopped
1 tablespoon salted capers (or more to taste)
2 small cans tomato sauce
2 ½ to 3 pounds *baccala*

Two days before Christmas, soak the *baccala* in cold water, changing the water at least two times a day. Cut *baccala* into pieces.

Sauté celery for five minutes. Add onion, garlic, and capers and cook a few minutes, until soft. Add tomato sauce. Cook over a low flame fifteen minutes. Add *baccala* and cook until fish comes apart with a fork.[28]

On New Year's Eve, the Baldizzis celebrated with *sfinge*, a kind of hole-less doughnut. Rosaria started the batter in the afternoon, mixing the flour, yeast, and water in a large pot, then covering it with a blanket and leaving it to rise. Just before midnight, she put a pot of oil on the stove, testing the temperature with a drop of water. When it splattered, the oil was sufficiently hot. The second the clock struck twelve, she dropped the first spoonful of batter, which caused the oil to bubble wildly, producing a loud *zhoosh*-ing sound. It took roughly a minute for the *sfinge* to cook up, puffy and golden. After scooping them from the pot, she dipped them in sugar. Then she fed them to the kids, still hot, so their first taste of the New Year would be sweet.

Notes

———

CHAPTER ONE: THE GLOCKNER FAMILY

1. Henriette Davidis, *Practical Cook Book: German National Cookery for American Kitchens* (Milwaukee, 1904), 131.

2. Davidis, *Practical*, 11.

3. Gesine Lemcke, "Cooking Correspondence," *Brooklyn Eagle*, January 1, 1899, 23.

4. Author's recipe, adapted from Davidis.

5. Davidis, *Practical*, 318.

6. Gesine Lemcke, "Cooking Correspondence," *Brooklyn Eagle*, March 26, 1899, 20.

7. "Uncleanly Markets," *New York Times*, May 22, 1854, 4.

8. "Market Reform," *New York Times*, March 29, 1872, 4.

9. "Local Intelligence," *New York Times*, December 19, 1865, 2.

10. Junius Henri Browne, *The Great Metropolis* (Hartford, 1869), 408.

11. Thomas F. De Voe, *The Market Assistant* (New York, 1862), title page.

12. "How New York Is Fed," *Scribner's Monthly*, October 1877, 730.

13. Mrs. Emma Ewing, *Salad and Salad Making* (Chicago, 1883), 37.

14. "Our City's Condition," *New York Times*, June 12, 1865, 1.

15. "Sauerkraut Statistics," *Chicago Tribune* (reprinted from the *Philadelphia News*), December 29, 1885, 5.

16. "The Sauerkraut Peddler," *Washington Post* (reprinted from the *New York Evening Post*), August 24, 1902, 10.

17. Charles Dawson Shanley, "Signs and Show-Cases of New York," *Atlantic Monthly*, May 1870, 528.

18. "Vienna Bread," *New York Times*, January 28, 1877, 6. ("Mackerelville" is a nineteenth-century term for the neighborhood that became the East Village.)

19. "The Household," *New York Times*, January 30, 1876, 9.

20. "Toothsome German Dishes," *New York Times*, July 11, 1897, 10.

21. "The Million's Beverage," *New York Times*, May 20, 1877, 10.

22. Browne, *Great Metropolis*, 161.

23. "History of Beer," *United States Magazine*, August 15, 1854, 180.

24. Jacob A. Riis, *How the Other Half Lives* (New York, 1890), 215.

25. "German Restaurants," *New York Times*, January 19, 1873, 5.

26. "Where Men May Dine Well," *New York Sun*, April 5, 1891, 23.

27. Emory Holloway, ed., *The Uncollected Poetry and Prose of Walt Whitman*, 2 vols. (Garden City, New York, 1921), II: 92.

28. Gesine Lemcke, *European and American Cuisine* (New York, 1933), 543.

29. "Luchow's," Benjamin DeCasseres, *American Mercury*, December 1931, 447.

30. "Germany in New York," *Atlantic Monthly*, May 1867, 557.

31. "The Yearly Turn-Fest," *New York Times*, August 26, 1862, 8.

32. "Jovial Souls," *Brooklyn Eagle*, July 14, 1891, 2.

33. "Sixth Plattdeustche Festival," *New York Times*, September 7, 1880, 8.

34. "New-York City. Germans in America," *New York Times*, June 27, 1855, 1.

Chapter two: The Moore Family

1. Andrew Carpenter, ed., *Verse from Eighteenth-Century Ireland* (Cork, Ireland, 1998), 248.

2. Nancy F. Cott, *Root of Bitterness: Documents of the Social History of American Women* (Lebanon, New Hampshire, 1996), 154.

3. Letter from P. Burdan, 1894. Personal Collection of Kirby Miller.

4. Letter from Cathy Greene, 1884. Personal Collection of Kirby Miller.

5. "Those Servant Girls," *Brooklyn Daily Eagle*, March 12, 1897, 3.

6. Johann Georg Kohl, *Ireland* (New York, 1844), 13.

7. Kohl, *Ireland*, 45.

8. Charles Loring Brace, *The Dangerous Classes of New York* (New York, 1872), 168.

9. Seamus MacManus, *Yourself and the Neighbours* (New York, 1914), 70.

10. Letter from Alice McDonald, 1868. Personal Collection of Kirby Miller.

11. Charles Fanning, *The Irish Voice in America* (Lexington, 2000), 127.

12. Louise Bolard More, *Wage-Earners' Budgets* (New York, 1907), 173.

13. "Cheap Pudding," *Irish Times*, February 1, 1879.

14. "A Day in Castle Garden," *Harper's New Monthly Magazine*, March 1871, 554.

15. John F. Maguire, *The Irish in America* (London, 1868), 190.

16. Jeremiah O'Donovan, *A Brief Account of the Author's Interview with His Countrymen* (Pittsburgh, 1864), 367.

17. "Saturday Night at Washington Market," *New York Times*, March 17, 1872, 5.

18. Maria Parloa, *First Principles of Household Management and Cookery* (Boston, 1879), 87.

19. "Restaurant Calls," *Brooklyn Daily Eagle*, July 3, 1887, 13.

20. J. C. Croly, *Jennie June's American Cookery Book* (New York, 1870), 76.

21. George Foster, *New York in Slices* (New York, 1850), 70.

22. William Ellis, *The Country Housewife's Family Companion* (Totnes, Devon, 2000), 97.

23. Kathleen Mathew, "New York Newsboys," *Frank Leslie's Popular Monthly*, April 1895, 458.

24. "Tempting Hotel Menus," *New York Times*, December 26, 1890, 8.

CHAPTER THREE: THE GUMPERTZ FAMILY

1. "Hester Street Market," *New York Times*, July 27, 1895, 12.

2. Ladies of Congregation Emanuel, *The Fair Cook Book* (Denver, 1888), 7. (Reproduced courtesy of the Beck Archives, Penrose Library, Special Collections, University of Denver.)

3. Leah W. Leonard, *Jewish Cookery, in Accordance with Jewish Dietary Laws* (New York, 1949), 166.

4. John Cooper, *Eat and Be Satisfied: a Social History of Jewish Food* (Northvale, New Jersey, 1993), 80.

5. Marx Rumpolt, *Ein new Kochbuch* (Frankfurt am Main, 1581), 120. Translated by the author.

6. Florence K. Greenbaum, *International Jewish Cookbook* (New York, 1919), 84.

7. "Where Strict Jews Eat," *Current Literature*, March 1881, 408.

8. Personal correspondence with Julia Kramer.

9. Bertha Kramer, *"Aunt Babette's" Cook Book* (New York, 1914), 513.

10. Matthew Hale Smith, *Sunshine and Shadow in New York* (Hartford, 1869), 456.

11. National Council of Jewish Women, *Council Cook Book* (San Francisco, 1909), 48.

12. Fannie Hurst, *The Vertical City* (New York, 1922), 262.

13. Albert Waldinger, ed., *Shining and Shadow: An Anthology of Early Yiddish Stories from the Lower East Side* (Cranbury, New Jersey, 2006), 142.

14. Michael Ginor et al., *Foie Gras, a Passion* (New York: 1999), 41.

15. Albert H. Buck, ed., *A Treatise of Hygiene and Health* (New York, 1977), 400.

16. During the early decades of the twentieth century, ethnically based food

rackets were common in New York City. Taking advantage of their fellow immigrants' fear and insularity, Jewish gangsters at various times took control of the kosher poultry industry, soda-fountain syrup manufacturing, and wholesale bakeries. Italian racketeers controlled artichokes, grapes, and pasta manufacture. All these rackets raised food prices mostly for that part of the population that could least afford it. Most food rackets were eliminated during the late 1930s, thanks to a concerted effort by the La Guardia administration.

17. "Some Queer East Side Vocations," *Current Opinion* (reprinted from the *New York Post*), August 1903, 202.

18. Greenbaum, *International Jewish Cookbook*, 12.

19. Author's family recipe.

20. Anya Yezierska, *Hungry Hearts* (New York, 1997), 116.

21. Henry Harlan, *The Yoke of the Thorah* (New York, 1896), 205.

22. Kramer, *"Aunt Babette's,"* 24.

23. Kela Nussbaum family recipe, contributed by Betsy Chanales.

CHAPTER FOUR: THE ROGARSHEVSKY FAMILY

1. "Humanity and Efficiency," *The Outlook*, March 28, 1908, 627.

2. Menu, Ellis Island archive.

3. "Their First Thanksgiving," *The Sun (New York)*, December 1, 1905, 2.

4. Frederick A. Wallis, "Treating Incoming Aliens as Human Beings," *Current History*, April–September 1921, 443.

5. "Feast of the Passover Celebrated," *New York Times*, March 26, 1899. 6.

6. Kosher menu, Ellis Island archive.

7. Frieda Schwartz family recipe, contributed by her daughter Francine E. Herbitter.

8. Regina Frishwasser, *Jewish American Cook Book* (New York, 1946), 47.

9. Bertha M. Wood, *Foods of the Foreign-Born in Relation to Health* (Boston, 1922), 90.

10. Jennie Grossinger, *The Art of Jewish Cooking* (New York, 1960), 147.

11. Elsa Herzfeld, *Family Monographs* (New York, 1905), 33.

12. Fannie Cohen family recipe, contributed by her granddaughter, Francine E. Herbitter.

13. Hinde Amchanitzki, *Text Book for Cooking and Baking* (New York, 1901), 30.

14. "East Siders Don't Approve Cookbook," *Hartford Courant*, January 17, 1916, 3.

15. John C. Gebhart, "Malnutrition and School Feeding," *Bulletin, United States Bureau of Education*, 1922, 14.

16. Emma Smedley, *The School Lunch* (Media, Pennsylvania, 1920), 147.

17. "Poor Meals Break Homes," *New York Times*, September 16, 1920, 8.

18. "Queer Dishes in Shops," *New York Tribune*, December 12, 1897, 40.

19. Alfred Kazin, *A Walker in the City* (New York, 1951), 34.

20. "Along Second Avenue," *New York Tribune*, August 31, 1919, F12.

21. Rian James, *Dining in New York* (New York, 1930), 32.

22. William Reiner, *Bohemia, the East Side Cafes of New York* (New York, 1903), 20.

23. Sholem Aleichem, *Wandering Star* (New York, 1952), 233.

24. "Along Second Avenue," *New York Tribune*, August 31, 1919, 68.

25. Family recipe of Lillian Chanales.

CHAPTER FIVE: THE BALDIZZI FAMILY

1. "Unskilled Laborer," *Washington Herald*, February 2, 1908, 12.

2. "New Recipe for Soup," *New York Times*, August 19, 1900, 15.

3. "Undesirable Immigrants," *New York Times*, December 18, 1880, 4.

4. "Phases of City Life," *New York Times*, November 4, 1871, 2.

5. "Italy's Invading Army," *New York Sun*, June 28, 1891, 23.

6. "Found in Garbage Boxes," *New York Times*, July 15, 1883, 10.

7. "Things Little Italy Eats," *The Sun (New York)*, August 23, 1903, 5.

8. Jerre Mangione, *Mount Allegro* (New York, 1981), 131.

9. See Hasia Diner's *Hungering for America*.

10. "Table Tidbits Prepared Under Revolting Conditions," *New York Tribune*, May 11, 1913, D4.

11. "Quaint Italian Customs of Summer Festal Days," *New York Times*, July 12, 1903, 30.

12. Maria Gentile, *The Italian Cook Book* (New York, 1919), 133.

13. Mangione, *Mount Allegro*, 131.

14. Richard Gambino, *Blood of My Blood: the Dilemma of the Italian American* (Toronto, 1996), 92.

15. Author's recipe.

16. Concetta Rizzolo's family recipe, contributed by her grandson Stephen Treffinger.

17. Jeannette Young Norton, "Going Marketing in 'Little Italy,'" *New York Tribune*, July 23, 1916, C5.

18. "Italian Housewives' Dishes," *New York Times*, June 7, 1903, 28.

19. "Do Fiery Foods Cause Fiery Natures?" *New York Tribune Illustrated Supplement*, December 6, 1903, 5.

20. Gentile, *The Italian Cook Book*, 76.

21. "A Bit of Bohemia," *The Vassar Miscellany*, February 1889, 154.

22. "March of the Italian Chef," *The Sun (New York)*, December 20, 1908, 8.
23. "Our Italians," *New York Times*, November 12, 1875, 4.
24. "A Sicilian Café in New York," *Harper's Weekly*, November 2, 1889, 875.
25. "The Italian Cook's Best," *The Sun (New York)*, June 20, 1909, 2.
26. Sarah Tyson Rorer, *Mrs. Rorer's New Cook Book* (Philadelphia, 1902), 301.
27. "Spaghetti with Meat Balls," *Chicago Tribune*, February 21, 1908, 9.
28. Baldizzi family recipe.

Bibliography

Barnavi, Eli, *A Historical Atlas of the Jewish People*. New York: Schocken, 1992.

Barr, Nancy Verde, *We Called It Macaroni: An American Heritage of Southern Italian Cooking*. New York: Alfred A. Knopf, 1990.

Bayor, Ronald H. and Timothy J. Meagher, eds., *The New York Irish*. Baltimore: Johns Hopkins University Press, 1996.

Beecher, Mrs. Henry Ward, *Motherly Talks with Young Housekeepers*. New York: J. B. Ford and Company, 1873.

Bolino, August Constantino, *The Ellis Island Source Book*. Washington, D.C.: Kensington Historical Press, 1985.

Browne, Junius Henri, *The Great Metropolis: A Mirror of New York*. Hartford: American Pub. Co., 1869.

Brownstone, David M. and Irene M. Franck, *Facts About American Immigration*. New York: H. W. Wilson Company, 2001.

Burrows, Edwin G. and Mike Wallace, *Gotham: A History of New York City to 1898*. New York: Oxford University Press, 1999.

Cahan, Abraham, *The Rise of David Levinsky*. New York: Harper & Brothers, 1917.

Clarkson, L. A., and E. Margaret Crawford, *Feast and Famine: Food and Nutrition in Ireland 1500–1920*. New York: Oxford University Press, 2001.

Coan, Peter Morton, *Ellis Island Interviews: In Their Own Words*. New York: Checkmark Books, 1997.

Cohen, Rose, *Out of the Shadow: A Russian Jewish Girlhood on the Lower East Side*. Ithaca, New York: Cornell University Press, 1995.

Cordasco, Francesco, ed., *Studies in Italian American Social History: Essays in Honor of Leonard Covello*. Totowa, New Jersey: Rowman & Littlefield, 1975.

Cowan, Cathal and Regina Sexton, *Ireland's Traditional Foods*. Dublin: Teagasc, 1997.

Dembinska, Maria, *Food and Drink in Medieval Poland: Discovering a Cuisine of the Past*. Philadelphia: University of Pennsylvania Press, 2002.

Diner, Hasia, *Hungering for America*. Cambridge, Massachusetts: Harvard University Press, 2001.

Dolkart, Andrew S., *Biography of a Tenement House in New York City: An Architectural History of 97 Orchard Street*. Santa Fe: The Center for American Places, 2006.

Ernst, Robert, *Immigrant Life in New York City, 1825–1863*. New York: King's Crown Press, 1949.

Ewen, Elizabeth, *Immigrant Women in the Land of Dollars: Life and Culture on the Lower East Side, 1890–1925*. New York: Monthly Review Press, 1985.

Fanning, Charles, *The Irish Voice in America: 250 Years of Irish-American Fiction*. Lexington, Kentucky: University Press of Kentucky, 2000.

Foner, Nancy, *From Ellis Island to JFK: New York's Two Great Waves of Immigration*. New Haven, Connecticut: Yale University Press, 2000.

Gabaccia, Donna R., *From Sicily to Elizabeth Street: Housing and Social Change Among Italian Immigrants, 1880–1930*. Albany, New York: State University of New York Press, 1984.

Gabaccia, Donna R., *We Are What We Eat: Ethnic Food and the Making of Americans*. Cambridge, Massachusetts: Harvard University Press, 1998.

Greenspoon, Leonard J., Ronald A. Simkins, Gerald Shapiro, eds., *Food and Judaism: A Special Issue of Studies in Jewish Civilization, Volume 15*. Omaha, Nebraska: Creighton University Press, 2005.

Hecker, Joel, *Mystical Bodies, Mystical Meals: Eating and Embodiment in Medieval Kabbalah*. Detroit, Michigan: Wayne State University Press, 2005.

Homberger, Eric, *The Historical Atlas of New York City*. New York: Henry Holt, 1994.

Hundert, Gershorn David, *The YIVO Encyclopedia of Jews in Eastern Europe*. New Haven, Connecticut: Yale University Press, 2008.

Isola, Antonia, *Simple Italian Cookery*. New York: Harper & Brothers, 1912.

Jackson, Kenneth, ed., *The Encyclopedia of New York City*. New Haven, Connecticut: Yale University Press, 1995.

Joselit, Jenna Weissman, *The Wonders of America: Reinventing Jewish Culture 1880–1950*. New York: Hill and Wang, 1994.

Joselit, Jenna Weissman, Barbara Kirshenblatt-Gimblett, Irving Howe, Susan L. Braunstein, *Getting Comfortable in New York: The American Jewish Home 1880–1950*. New York: The Jewish Museum, 1990.

Kagan, Berl, ed., *Luboml: The Memorial Book of a Vanished Shtetl*. Jerusalem: Ktav Publishing House, 1987.

Kessner, Thomas, *The Golden Door: Italian and Jewish Immigrant Mobility in New York City 1880–1915*. New York: Oxford University Press, 1977.

Maffi, Mario, *Gateway to the Promised Land: Ethnic Cultures in New York's Lower East Side*. New York: New York University Press, 1995.

Maguire, John Francis, *The Irish in America*. New York: D. & J. Sadlier & Co., 1868.

Mangione, Jerre and Ben Morreale, *La Storia: Five Centuries of the Italian-American Experience*. New York: HarperCollins, 1992.

McCabe, James D., *Lights and Shadows of New York Life*. Philadelphia: National Publishing Company, 1872.

Moreno, Barry, *Encyclopedia of Ellis Island*. Westport, Connecticut: Greenwood Press, 2004.

Nadel, Stanley, *Little Germany: Ethnicity, Religion, and Class in New York City, 1845–80*. Urbana, Illinois: University of Illinois Press, 1990.

Nathan, Joan, *Jewish Cooking in America*. New York: Alfred A. Knopf, 1998.

Parker, Cornelia, *Working with the Working Woman*. New York: Harper & Brothers, 1922.

Parloa, Maria, *First Principles of Household Management and Cookery*. Boston: Houghton, Osgood and Company, 1879.

Pitkin, Thomas, *Keepers of the Gate: A History of Ellis Island*. New York: New York University Press, 1975.

Plunz, Richard, *A History of Housing in New York City*. New York: Columbia University Press, 1990.

Rischin, Moses, *The Promised City: New York's Jews, 1870–1914*. New York: Corinth Books, 1964.

Roden, Claudia, *The Book of Jewish Food*. New York: Alfred A. Knopf, 1996.

Roskolenko, Harry, *The Time That Was Then*. New York: Dial Press, 1971.

Sarna, Jonathan D., and Nancy H. Klein, *The Jews of Cincinnati*. Cincinnati: Center for the Study of the American Jewish Experience, 1989.

Schneider, Dorothee, *Trade Unions and Community: The German Working Class in New York City, 1870–1900*. Urbana, Illinois: University of Illinois Press, 1994.

Simeti, Mary Taylor, *Pomp and Sustenance: Twenty-Five Centuries of Sicilian Food*. New York: Alfred A. Knopf, 1989.

Strasser, Susan, *Never Done: A History of American Housework*. New York: Pantheon Books, 1982.

Ware, Caroline Farrar, *Greenwich Village, 1920–1930: A Comment of American Civilization in the Post-War Years*. Berkeley: University of California Press, 1994.

Photo Credits

Index